Charles Seale-Hayne Library
University of Plymouth
(01752) 588 588
LibraryandITenquiries@plymouth.ac.uk

Fracture and Fatigue in Wood

Fracture and Fatigue in Wood

Ian Smith

*Faculty of Forestry and Environmental Management,
University of New Brunswick, Canada*

Eric Landis

*Department of Civil and Environmental Engineering,
University of Maine, USA*

Meng Gong

*Wood Research Institute,
Kyoto University, Japan*

WILEY

Copyright © 2003   John Wiley & Sons Ltd, The Atrium, Southern Gate, Chichester,
West Sussex PO19 8SQ, England

Telephone (+44) 1243 779777

Email (for orders and customer service enquiries): cs-books@wiley.co.uk
Visit our Home Page on www.wileyeurope.com or www.wiley.com

This publication is designed to provide accurate and authoritative information in regard to the
subject matter covered. It is sold on the understanding that the Publisher is not engaged in rendering
professional services. If professional advice or other expert assistance is required, the services of a
competent professional should be sought.

Other Wiley Editorial Offices

John Wiley & Sons Inc., 111 River Street, Hoboken, NJ 07030, USA

Jossey-Bass, 989 Market Street, San Francisco, CA 94103-1741, USA

Wiley-VCH Verlag GmbH, Boschstr. 12, D-69469 Weinheim, Germany

John Wiley & Sons Australia Ltd, 33 Park Road, Milton, Queensland 4064, Australia

John Wiley & Sons (Asia) Pte Ltd, 2 Clementi Loop #02-01, Jin Xing Distripark, Singapore 129809

John Wiley & Sons Canada Ltd, 22 Worcester Road, Etobicoke, Ontario, Canada M9W 1L1

Wiley also publishes its books in a variety of electronic formats. Some content that appears
in print may not be available in electronic books.

British Library Cataloguing in Publication Data

A catalogue record for this book is available from the British Library

ISBN 0-471-48708-2

Contents

1
Introduction

1.1 General Background

Wood in its unmodified form is unique amongst commercially important engineering materials because it is a product 'designed' and produced by nature to meet the needs of trees, rather than being something artificial that man makes from raw constituents. It is extremely mechanically efficient compared to most other materials, and withstands damaging effects in the growth environment and then damaging effects of unnatural environments after trees have been harvested and processed into components. Variability and heterogeneity are inherent traits of the material. The structure of wood, at any level one cares to visualise it, is highly optimised according to needs of trees. Wood is a natural composite with cellular structure that results in strongly directional dependent properties. The types and arrangements of cells are those that facilitate transportation of water and nutrients within trees, and that combat mechanical damage to the stems and branches by loads such as those due to wind, snow and ice. Different species have different structures because they exploit different environments and the same environments in different ways. Reflecting that diversity is the key to survival.

Water is essential to any tree's functioning. The amount of moisture held within wood can drastically influence both physical and mechanical properties and they need to be indexed to an associated moisture state. Stiffness and strength tend to reduce with any increase in the moisture content, but only up to the moisture level at which cell walls become saturated. If it were not for the cell walls being saturated, standing trees would be excessively stiff and snap instead of swaying in the wind. When man uses wood it is usually dried to well below the level where cell walls are saturated. Although this stiffens and strengthens the material, it also makes it more brittle, i.e. prone to fracture.

In a living tree, wood at any particular location experiences temporal variation in stress due to growth processes and environmental loads. Some damage is unavoidable and can appear at the cellular or gross level. Cell level damage is impossible to detect without using very high level magnification. Harvesting, processing, transportation and handling can cause further damage. Wood components are very unlikely to be undamaged when they enter service. In service wood components accumulate further damage

Fracture and Fatigue in Wood   I. Smith, E. Landis and M. Gong
© 2003 John Wiley & Sons, Ltd   ISBN: 0-471-48708-2 (HB)

due to excessive external load, moisture movement, excessive heat, decay and insect attack. Some of these factors have negligible influence for objects that are designed, fabricated, used and maintained properly. However, it is rarely economic to design against all damaging possibilities. Even trees recognise this, which is why some break prior to their natural mortality. Engineers conventionally regard wood components as being completely damage free prior to their end use application. Although convenient, this is a rather dubious presumption. Pre-service, often called inherent damage is responsible for a large proportion of intrinsic variability in engineering properties and life expectancy of wood components.

Generally, wood does not yield or flow under high levels of stress in a manner analogous to metals. There are no theories for wood comparable to those of yield, energy of distortion, flow and hardening. Nevertheless, people who study wood quite often borrow theories for other materials. Not surprisingly, this tends to be an unproductive avenue. Much information about failure processes in wood was acquired empirically in the early 1900s for aircraft engineering purposes, and more recently for civil engineering purposes. This tends to be reported in the form of empirical cause and effect relationships. This essentially black box approach has supported utilisation of wood, and is the basis of much of contemporary engineering design. As with any other black box techniques, gaps in data are routinely filled by judgement. Filling gaps is especially difficult when it comes to extrapolation rather than interpolation. Both interpolation and extrapolation procedures are only reliable if they are based on detailed knowledge of behavioural mechanisms, i.e. the knowledge that black box approaches side step is needed. This does not mean that black box methods need to be abandoned altogether. What is required is a more balanced and pervasive development of understanding. In the context of wood, links need to be established between materials science/wood science, mechanics and engineering knowledge. This is the basic premise that motivated the writing of this book.

Inherent damage in woods usually equals cracks of various sizes. Analysis of cracked bodies equals fracture mechanics. Progressive growth of damage equals fatigue. Also, wood and other materials with strongly aligned cellular structures are prone to development of intense stress regions (stress concentrations) at the local or higher level. Stress concentrations promote fracturing, i.e. cracks. Whatever way the question is looked at, it seems obvious that a deep understanding of fracture and fatigue mechanisms, and associated theories, is key to understanding the mechanical behaviour of wood. Hence this book is needed.

1.2   Why Apply Fracture Mechanics Principles to Wood?

The question can be answered illustratively via consideration of a phenomenon that is only rationally explained through fracture mechanics principles. There exist so-called volume effects on the apparent strength of wood members, whereby strength changes as the volume changes. Under most stress conditions apparent strength reduces as member volume increases. This phenomenon is usually attributed to statistical variability in material properties and is simulated using weakest link theories. The basis of such theory is that properties and inherent flaws in wood are homogenously dispersed both within and between members, with the consequence that the likelihood of a high stress coinciding with a weak location somewhere within a member will increases with any increase in member dimensions. It is further assumed that any local failure (stress > strength) will invariantly develop in an unstable manner to cause catastrophic global failure. Unfortunately, this logic must be flawed, because even the most casual perusal reveals that neither properties nor flaws in wood can sensibly be regarded as homogeneously dispersed. The true explanation of volume effects lies elsewhere.

Inherent damage in wood is in the form of cracks. It is created in the living tree or subsequent to their harvest as already mentioned. Propagation of a particular flaw (crack) amounts to providing enough energy to extend its length. The energy required for crack extension depends on its size and not on the size of member in which it is located. Suppose that wood members of various sizes are tested to determine their strengths under displacement control loading, and that stems from which members are derived could have been used to produce members of different size. As already implied, sizes of particular flaws are unrelated to sizes of member in which they end up being located. If a crack in a member extends, energy that needed to extend it can only come from reducing the strain energy stored in the member as a whole. Strain energy stored in a member is proportional to its volume and the load level. When a crack in a small member extends by a finite amount, there can be a significant release of strain energy from the member and measurable increase in its compliance. Crack growth can often be arrested instead of propagating in an unstable manner, because the wood is heterogeneous and because the strain energy available to drive further crack extension is reduced. By contrast, where the same crack to be located within a much larger member, there would be no appreciable change in the total strain energy, or compliance of the member, if the crack grew by a finite amount. If there is no appreciable change in the strain energy, i.e. no change in the potential for crack extension, it follows that conditions for crack growth persist. Even in the presence of material heterogeneity, growth of pre-existing cracks in large members is unlikely to be arrested and will quickly lead to catastrophic failure of such members.

Although the explanation above is somewhat simplified, it clearly illustrates that there are situations for which fracture mechanics-based explanation of phenomenon are highly plausible. The advantage of fracture mechanics explanations is that they are based on physical phenomena. Consequently, they should not break down when applied beyond the range of calibration data, as could alternative, apparently serendipitous (but flawed) explanations.

1.3 Why Apply Knowledge of Fatigue Processes to Wood?

As in the previous section, the question will be answered via consideration of a behaviour that can only be explained through understanding and modelling of the physical phenomenon. In whatever region of the world one lives, there will be familiarity with cases where there was an unexpected and superficially unexplainable structural failure. Consider a hypothetical roof that withstood the effects of heavy snowfalls over a number of years, but which suddenly collapsed during a similar event of lesser intensity. How is this explained? Suppose that the roof has wood trusses that entered service without any visible cracks and with quite high moisture content (>19%). The lower chords of the trusses are exposed to the heated internal climate of the building. Irrespective of what was visible, the wood in trusses had inherent damage in the form of micro cracks lying in the parallel to grain direction. Once in service, the bottom chord members started to dry, and over a period of several years moisture content stabilises to about 6% in summer and 8% in winter. Both the initial drying and seasonal fluctuations in moisture created moisture gradients, and therefore stress in the wood. Most importantly, tensile surface stress developed perpendicular to grain causing some regions of excessive stress at crack tips and normal to crack planes. This caused existing micro-cracks to propagate and some new micro-cracks to form. Even though some of the cracks coalesced, the main consequence was release of moisture-induced stress, and they did not lead to unstable major cracking. Concurrently, stress was generated in the wood due to self (dead) weight of the roof structure and environmental loads on roof surfaces. For the particular roof, peak external load induced stresses occurred when snow was on the roof, with such events lasting several days or weeks at a time. As far as the pre-existing cracks in the wood were concerned, the snow loading caused excessive shear stress at the tips parallel to crack planes. This caused the cracks to slowly grow in length (static fatigue) during each snow event. Each time cracks grew whether due to moisture movement or external loading, the residual strength of the wood and therefore the trusses and the roof reduced by some amount. It is clear that damage (crack lengths) increased due to multiple causes that interacted nonlinearly, even though they stressed crack tips differently. By how much the strength was reduced depended upon the extent of pre-existing damage and intensity of stress causing events (drying and external loading). During the snowfall prior to structure collapse, the severity of the damage (length of cracks) reached significant levels, but the structure survived because the event terminated just before the damage could propagate in an unstable manner. When the next event came along the roof collapsed, even though the loading was less intense in terms of the amount of snow than before. Failure occurred because the last loading was sustained long enough for slow crack growth to become unstable.

The sequence of events just described is a fatigue process. With advance knowledge of such processes, it would have been possible to predict the likely level of damage that snow load pulses and moisture effects would inflict on the wood trusses. Truss spans could have been limited to a length that kept stresses low enough that failure was much less unlikely.

1.4   Arrangement of this Book

The approach of this book is to bring together all information pertinent to a proper acquaintance with wood mechanics, fracture and fatigue processes, and prediction of the same. Emphasis is on principles, rather than giving 'cook book' answers. Discussion starts with trees and wood properties, moves to fracture mechanics, then to findings of experimental studies into fracture and fatigue in wood, then to modelling fracture and fatigue in wood, and finally, to some practical applications of information and concepts. Both wood science and engineering perspectives are brought to bear, and hopefully this leads to a holistic treatment.

As will be apparent from the above discussion, fracture and fatigue are related physical processes, with both being associated with crack propagation. Strictly, fatigue behaviour is simply one manifestation of fracture behaviour, and attempts have been made at unified explanations and modelling. Normally, however, there are quite distinct test methods, analytical methods and literatures associated with fracture and fatigue.

Layout of subsequent chapters largely reflects and respects that tradition, not least because it keeps concepts manageable and facilitates explanation. That should not detract from value of the book as long as readers are aware that the separation is essentially artificial.

The book is standalone in the sense that it should not be necessary for readers with an engineering or materials science background to refer to other texts in order to follow discussion on fracture and fatigue in wood. Chapters are modular and are also standalone units. This is to recognise that some readers will feel themselves adequately knowledgeable about certain content that they wish to skip ahead. There is extensive cross-referencing between chapters to ensure that those who do this will not lose the thread of continuing arguments. The scopes of subsequent chapters are as follows.

Chapter 2 discusses the structure and cellular nature of wood, and its physical and mechanical properties. Consideration ranges from the molecular level to the gross level. Emphasis is placed on the material's biological origin, what influences growth processes and structural features, and how and why wood can be damaged prior to its end-use application. Discussion of mechanical properties is fairly brief, and focuses on the orders of magnitude and why they vary within one tree stem, within a species and between species. Information given is that necessary for full understanding of later discussion about dependence of fracture and fatigue properties on factors such as species, density and moisture content.

Chapter 3 discusses conceptualisation and modelling of mechanical properties of wood at various levels of detail. There is no universally appropriate level at which to model wood, as that depends upon an analyst's purpose and the type of problem being solved. The chapter starts with models for properties on the micro scale, and eventually arrives at representation of wood on the massive scale. Emphasis is placed on the macro scale (wood free of visible defects), because this is the level often appropriate to numerical fracture analysis. Attention is drawn to the need to recognise that so-called mechanical properties of wood are only constant under well-defined conditions. Ignoring this can lead to poor modelling and misapplication of the material.

Chapter 4 sets the scene for later chapters by discussing the principles of fracture mechanics. Its place is as an alternative to traditional strength theories that prove unable to correctly predict the strength of brittle materials. Fracture mechanics theories recognise that cracks and flaws in materials ultimately dictate fracture strength. Such theories focused on the role of a single critical crack, and their development is traced to its roots. The story starts with theories that assume linear elastic material response. Importance of various toughening mechanisms is highlighted, which renders linear elastic methods inappropriate for many applications of some materials. Various alternative theories have been developed because of this. Damage mechanics is a field that deals with distributions of flaws and their cumulative effect on mechanical behaviour, and as this is often the case with wood; that topic is also introduced. Although not all theories discussed are suitable for wood, it is important that they all be understood as a basis for choosing those that are.

Chapter 5 presents and interprets experimental information about the fracture behaviour of wood. Discussion focuses on the relationship between microstructural changes that occur as a result of fracture and the measured bulk fracture properties. Different fracture and failure modes are classified relative to the growth axis of the wood, with special attention paid to tension perpendicular to the grain. Most of the past and current developments in fracture mechanics of wood have been in this area. These developments are presented along with limitations to current knowledge and possible avenues for future improvements.

Chapter 6 presents and interprets experimental information about cyclic and static fatigue behaviour of wood. The focus is on the behaviour of wood without visible defects, massive (structural) wood members, and 'wood with glue' composites. Attention is drawn to how the heterogeneous and rheological nature of such materials influences the process by which they accumulate damage. Most externally applied loads on wood components are cyclic in nature, and it is important to recognise that damage in wood depends upon the number of stress cycles and the rate of stressing, as well as time under load. Only the latter is commonly considered, which is a serious omission.

Chapter 7 discusses models for predicting fracture in wood. Strength predictions based on Weibull theory are presented as a traditional approach to account for size effects. The shortcomings of the theory are rooted in its purely statistical underpinnings. Deterministic modelling efforts based on a 'finite element' framework are presented as a way to incorporate fracture mechanics concepts into practical engineering design and analysis. Continuum-based finite element methods are not without serious limitation, however. While extremely computationally efficient, the method becomes laced with empirical constants when damage and fracture are introduced into the continuum. A computational approach based on a lattice representation of the material is presented as a potential alternative to continuum-based approaches.

Chapter 8 discusses models for predicting cyclic and static fatigue in wood, components and structural systems. Although phenomenological and mechanics based models have been proposed, most models are empirical. However, irrespective of their basis and level of complexity, models require extensive experimental calibration. It is argued that currently there is no universally superior model, and choice of a best model depends upon an analyst's purpose. Attention is drawn to issues such as whether models need to incorporate a threshold stress below which stress induced damage does not accumulate, because predictions from certain models are highly sensitive to this choice.

Chapter 9 presents a series of practical problems to illustrate the application of information and concepts of fracture and fatigue in wood. Although these are not all encompassing, they demonstrate what is possible with current knowledge. Specific problems deal with fracture of members and joints, and low cycle and high cycle fatigue of members and joints. Underpinning them is a presumption that information and concepts should be applied as a heuristic process, with the most important ingredient for success being deep understanding of underlying processes.

2
Structure and Properties of Wood

2.1 The Nature of Wood

Wood is a natural, heterogeneous, anisotropic, hygroscopic composite material with high strength and stiffness relative to its weight. Unlike other common structural materials, it often exhibits a markedly viscous response to external influences under normal use conditions. As the famous American architect Frank Lloyd Wright said in 1928, "We may use wood with intelligence only if we understand wood". The beginning for such understanding is knowledge of how trees grow and how they have engineered their substance so it has traits advantageous to their existence.[1] One cannot avoid starting in the forest, so to speak. Knowledge of wood has been gained following the deconstructionist (scientific) approach, starting with gross features and behaviour, and has moved downward in scale through features and functions of constituent parts, with molecular composition being the lowest rung on the ladder (Figure 2.1). This way of gaining insight contrasts with the integrationist (engineering) approach, which is the basic tenet of this book, i.e. building from functions and behaviour of parts to behaviour of any system. It is hoped readers find juxtaposition of these approaches an interesting counterpoint within the overall discourse.

2.1.1 Tree growth and wood formation

Trees are perennial, vascular and woody plants having two principle groups in terms of their anatomy. One is called conifers, evergreens or softwoods, which commonly have needle-like leaves and naked seeds. The other is known as broad-leaved trees, deciduous trees or hardwoods, which have broad leaves and enclosed seeds. Strictly speaking, softwoods and hardwoods are terms for commercial purposes used with no intention of implying they actually are relatively hard or soft. In fact, both the hardest

---

[1] Although in relatively modern times man has intervened in their evolutionary process through genetic manipulation, commercially produced wood is by and large an entirely natural material.

Fracture and Fatigue in Wood I. Smith, E. Landis and M. Gong
© 2003 John Wiley & Sons, Ltd ISBN: 0-471-48708-2 (HB)

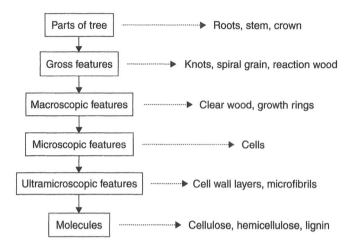

Figure 2.1   Levels at which wood structure is classified and analysed.

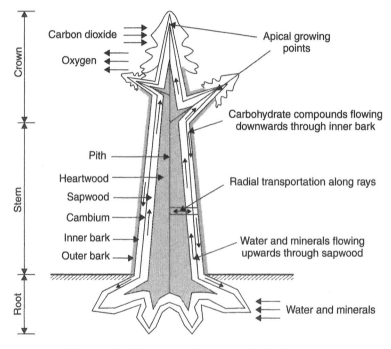

Figure 2.2   Main parts of a tree and their functions.

and softest woods on the earth are hardwoods (see Section 2.3.1). However, hardwoods are usually denser and therefore harder than softwoods.

Trees contain three main parts: root system, stem, and crown system (Figure 2.2), with each part being multi-functional. Roots anchor a tree to the ground, support it, and absorb water and minerals. Usually, the stem(s) contains most tree substance above the ground, it conducts sap (mixture of water and minerals) from roots to leaves through sapwood, stores food materials, and supports the crown. The crown system contains branches, twigs and leaves. Leaves absorb carbon dioxide from the atmosphere, and capture energy from sunlight, then combine them with sap by photosynthesis to manufacture food materials (carbohydrate compounds). Food materials travel downwards and are distributed through the inner bark of the stem. Structural, transportation and storage functions in a tree are performed by cells, with aggregations of cells of similar type, or similar function, being known as tissues.

Figure 2.3 illustrates the principle components in the cross-section of a stem. The pith is the primary tissue in the form of a central parenchymatous cylinder. Within the darker coloured portion called heartwood that surrounds the pith, cells are dead even in a live tree. Cells in the outer lighter coloured portion called sapwood surrounding the heartwood, are alive in a live tree. Together heartwood and sapwood constitute the xylem that is synonymous with 'wood'. All cells forming wood are produced by the vascular cambium, as in any woody plant. The cambium is a narrow layer of cells located between the inner bark and xylem. Outside of the cambium is bark composed of an inner living bark and outer dead bark. As can be seen in Figure 2.3, although the cross-section is approximately circular, trees do not necessarily develop uniformly in all radial directions due to influences such as direction of sunlight, shade and topography of the growth site.

The stem is of greatest interest to wood using industries, and accounts for between about 50% and 90% of the whole tree by volume for commercial species. The size of stem increases as a tree ages until it dies naturally, due to disease or disasters, or is harvested to serve human needs. Growth of a tree is an additive and cumulative

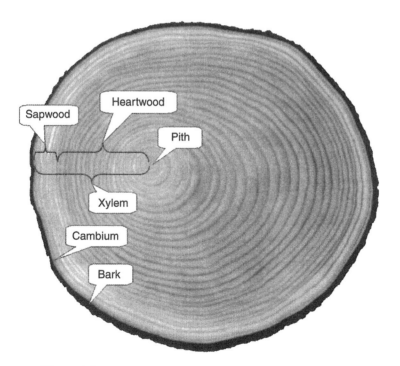

Figure 2.3 Cross-section of a Tamarack (*Larix laricina*) stem.

process that involves longitudinal and radial growth of stem, branches and twigs. Longitudinal growth of a stem and branches originates from the primary growth that takes place at, or near, the apical growing points (Figure 2.2). Radial growth is viewed as secondary growth of a stem and branches and is the result of cambial activities. During a growth season, the cambium produces a new sheath of wood inwards, and additional inner bark outwards. Trees optimise themselves over their growth, usually resulting in an upward taper of the stem, which develops its greatest girth at the base in order to support the weight of a tree and possibly snow and ice, and to resist wind loading. Trees of the same species can have quite varied characteristics. Features, such as the amount of taper, depend upon growth conditions as well as genetics.

During the growth of a tree, internal stresses develop in the stem. Such internal stresses (so called growth stresses) initiate in the cambium as it adds each new layer of wood to the stem. There is no consensus about exactly how growth stresses are generated. The two main hypotheses are the lignin swelling hypothesis (Munch, 1938) and the cellulose tension hypothesis (Bamber, 1978). The former is that the deposition of encrusting lignin between cellulose fibrils causes transverse expansion of wood cells. The latter is that tensile stresses arise from contraction of cellulose during crystallisation. Growth stresses consist of longitudinal tensile stresses in the outer layers of a stem and compressive stresses in the inner core. They usually disappear quickly after felling (Mattheck and Breloer, 1994). Maximum tensile stresses of about 10 MPa are reported in large diameter European hardwoods, and over 14 MPa in Australian eucalypts (Alhasani, 1999). The stem is dominated by tensile growth stress in the radial direction, while in the tangential direction, the growth stress changes from compression at the periphery to tension near the pith (Alhasani, 1999). Figure 2.4 illustrates the distribution of internal stresses within a stem.

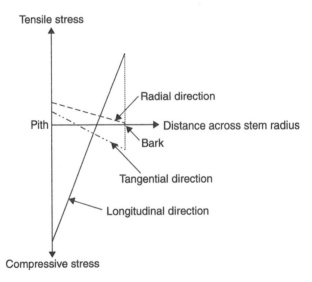

Figure 2.4   Schematic distribution of growth stresses in three directions.

2.1.2 Gross growth features in wood

Trees have quite varied growth features within stems and branches, with some features being exhibited by all trees and others depending on species/genetics and growth conditions. This section discusses various features associated with variability in structure and properties of wood.

Juvenile wood is a sheath of wood surrounding the pith and typically varies from 5–20 growth rings in radius. It represents xylem cells formed by juvenile or immature cambium and occurs in all trees. The top of a tree stem is predominately juvenile wood, but the top of some species, e.g. larch, includes mature wood. Juvenile wood is usually more predominant in softwood than in hardwoods. Compared with mature wood (formed by mature cambium), it is characterised by short cells with large diameter lumen, thin cell walls, and tends to have more knots. Within the juvenile region of the xylem, physical and mechanical properties, including cell size, vary. Mature wood has relatively constant cell dimensions and stable properties.

Knots are parts of tree branches that are embedded in the main stem or other branches. Lateral branches are connected to the pith of the main stems or other branches. The cambium of a branch is continuous with that of a stem while the branch is alive, resulting in the formation of intergrown or tight knots. Once a branch dies, the continuity is broken, producing an encased or loose knot. Knots create localised disturbances in cell orientations (grain disturbances), resulting in a change in wood structure and therefore wood properties. Whorls of lateral branches occur at regular intervals in many softwood species resulting in knot clusters in lumber products.

Reaction wood is an abnormality formed in leaning trees or branches and is associated with eccentrically located pith and abnormal growth rings. The formation of reaction wood is thought to be a mechanism for bringing a leaning stem back to an upright position, or for maintaining a proper orientation of a branch. Reaction wood differs considerably in softwoods and hardwoods. In softwoods, the growth rings are abnormally wide on the underside of a leaning stem or branch leading to what is called compression wood. While in hardwoods, the thickening occurs on the topside of a leaning stem or branch leading to so called tension wood. This reflects that trees can optimise their strengthening processes differently, which in turn reflects mechanical capabilities of their cell structures. Key disadvantages of reaction wood are reduced strength and increased shrinkage along the grain, compared to normal wood.

Another important growth feature is spiral grain, in which the longitudinal cells are wound around the stem in a helix, giving a twisted appearance to the stem after removal of the bark. It is believed to be a result of the divisions of cambial cells. Although spiral grain is primarily a hereditary trait, environmental conditions can play an important role. One reasonable explanation of the origin of spiral grain is that trees do everything they can to minimise influences of external loads and avoid stress concentrations (Mattheck and Breloer, 1994). Spiral grain may follow either right- or left-handed patterns,[2] but it is seldom consistent through the stem radius. There are basically two patterns in softwoods. One pattern is a left-handed spiral that

---

[2] Right-handed spiral means that the deviation of grain is to the right of the longitudinal axis of a tree, and left-handed means that the deviation is to the left, from the butt to the top.

starts near the pith and develops over about ten juvenile growth rings, then gradually changes to a straight grain condition, and finally switches to a right-handed spiral pattern with the slope increasing towards the cambium. The other pattern is a left-handed spiral that keeps developing from the pith to the cambium. Spiral patterns in hardwoods are more complex than those in softwoods, and do not have fixed modes.

Although it is generally accepted that features such as juvenile wood, reaction wood, knots and spiral grain adversely influence physical and mechanical properties, this is not always the case.

2.1.3 Chemical components

Like other biological material, wood is a carbohydrate, the basic chemical elements of which are carbon, hydrogen and oxygen in the ratios $1:2:1$. These three elements form the primary substances of wood, namely cellulose, hemicellulose and lignin, and account for 97–99% of oven-dry wood substance by weight. In addition, there exist some organic materials known as extractives such as tannins, resins and gums, and less than 1% inorganic ash material.

Cellulose is a long linear unbranched chain polymer made up of 5000–10 000 glucose units and is organised into slender microfibril strands with periodic crystalline and non-crystalline regions along their length. The greater the length of the polymeric chain, the higher the degree of polymerisation, hence the greater the strength of the cell wall and the strength of the wood. Crystalline regions are areas in which cellulose units are arranged parallel to each other in highly ordered chains, which do not absorb water. Non-crystalline regions are areas in which cellulose is relatively disordered, which can absorb moisture and alter physio-mechanical properties of a microfibril. This explains why properties of wood can vary with changes in relative humidity and temperature even under normal use conditions. Cellulose accounts for 40–50% of wood substances by weight and has a density of about 1550 kg/m^3.

Hemicellulose has linear and highly branched chain backbones with a lower degree of polymerisation than cellulose. Some hemicellulose is amorphous and some oriented. Hydrogen bonding exists both within hemicellulose chains and between them and cellulose chains. Hemicellulose exhibits high moisture adsorption capacity. It accounts for about 25% of wood substance by weight in softwoods, and about 35% in hardwoods, and has a density of about 1500 kg/m^3.

Lignin is completely amorphous and has a three-dimensional molecular structure in which phenylpropane constitutes the basic skeleton. It accounts for about 20–30% of wood substances by weight in softwoods. Generally hardwoods have slightly less lignin. Lignin is the most hydrophobic component in the cell walls and has a density of about 1400 kg/m^3. Lignin is essentially the adhesive that binds other components together. It begins to soften at about 170°C, which is a characteristic exploited in the manufacture of certain wood-based products.

The average density of wood substance is almost invariant at 1500 kg/m^3. As is discussed in Section 3.2, wood's chemical composition, and its volumetric proportions and the disposition of the components leads directly to an ability to model physical and mechanical properties of cell walls, and hence ability to model properties of wood in bulk.

2.1.4 Microfibrils and cell walls

The basic structural element of cell walls is bundles of cellulose in association with hemicellulose and lignin, known as microfibrils. Microfibrils have a rectangular cross-section 3.5×10 nm and are of indefinite length. There exists a series of crystalline and non-crystalline regions along the length. Orientation of microfibrils in relation to the longitudinal axis of a cell is known as the microfibril angle (*MFA*). The *MFA* influences the anisotropic nature of cell walls and is ultimately reflected in the mechanical behaviour of wood (see Section 3.2).

A cell consists of a wall and a lumen, with the wall having a series of layers (Figure 2.5). All wall layers except the middle lamella (M) consist of microfibrils, but differ in the orientation of the microfibrils, and the thickness and proportions of chemical components. The extremely thin primary wall (P) is usually associated with the middle lamella that bonds adjacent cells together. Generally, the middle lamella plus primary walls from two neighbouring cells constitute a compound middle lamella. The primary wall has loosely packed and interwoven microfibrils without lamellate structure. The secondary wall (S) consists of microfibrils that are closely packed and parallel to each other. It can be subdivided into outer (S_1), middle (S_2), and inner (S_3) layers. As implied by Table 2.1, the S_2 layer is the dominant influence on some wood properties. Reflecting this, the *MFA* of the S_2 layer is that quoted in discussion on wood mechanics.

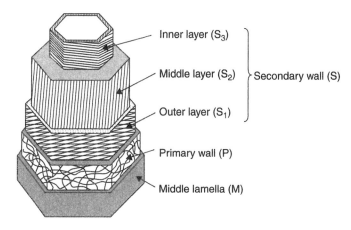

Inner layer (S_3)

Middle layer (S_2) ⎫ Secondary wall (S)

Outer layer (S_1) ⎭

Primary wall (P)

Middle lamella (M)

Figure 2.5 Structure of cell wall.

Table 2.1 Microfibril angles and relative thickness of the cell wall layers in spruce (based on Schniewind (1989) and Dinwoodie (1989))

Wall layer	Relative thickness (%)	Microfibril angle (*MFA*)	Number of lamellae
M + P	5	Random	None
S_1	10	50°–70°	4–6
S_2	75	10°–30°	30–150
S_3	10	60°–90°	0–6

Viewed from a materials science perspective, wood is a two-phase material with layers in cell walls consisting of crystalline cellulose forming fibres in a matrix of non-crystalline cellulose, hemicellulose and lignin, with minor amounts of extraneous materials.

2.1.5   *Microstructure*

As previously mentioned, wood is an aggregation of cells, the shapes of which are basically fibre-like and non-fibre-like (Figure 2.6). The fibre-like cells are aligned in essentially parallel to the pith (longitudinal) direction of a stem or branch, have relatively thick cell walls, have large length-to-width ratios, and normally play a mechanical role. These cells encompass tracheids in softwoods, and fibres in hardwoods. The non-fibre-like cells usually appear in series forming tissues, are oriented longitudinally and radially relative to the pith, have relatively thin or thick cell walls, and perform transportation and storage functions. These cells or tissues encompass longitudinal parenchyma, ray parenchyma and vessels. There are no components oriented in the tangential to a stem or branch direction. Arrangements of cells vary from species to species, but they always result in different distributions of cells in longitudinal, radial and tangential directions. Primary functions of various kinds of cell are given in Table 2.2.

Most softwood species do not have more than four or five types of cells, and only longitudinal tracheids and ray parenchyma occur in appreciable numbers. Therefore,

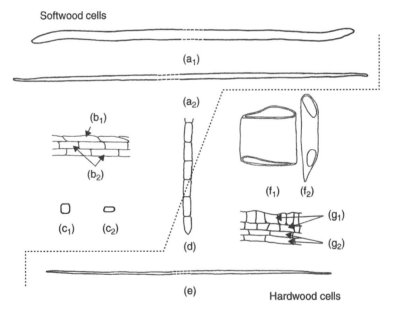

Figure 2.6   Cells in wood: (a_1)-Tracheid in earlywood, (a_2)-Tracheid in latewood, (b_1)-Ray tracheids, (b_2)-Ray parenchyma cells, (c_1)-Epithelial cell in vertical resin canal, (c_2)-Epithelial cell in horizontal resin canal, (d) Longitudinal parenchyma cells, (e) Fibre, (f_1)-Vessel element in earlywood, (f_2)-Vessel element in latewood, (g_1)-Upright ray cells, (g_2)-Procumbent ray cells.

Table 2.2 Main function of wood cells

Function	Softwood	Hardwood
Mechanics	Tracheids	Fibres
Transportation	Tracheids	Vessels
	Ray tracheids	Ray parenchyma
	Ray parenchyma	
Storage	Longitudinal parenchyma	Longitudinal parenchyma
Secretion	Epithelial cells	Epithelial cells

the structure of softwood is quite simple and uniform with 90–95% of volume comprised of long tracheids 25–45 μm in diameter and 3–5 mm in length. Properties of softwood are mainly determined by tracheids. The most obvious feature of softwood is its radial alignment in the cross-section, which means that softwood is strongly anisotropic in the cross-section plane (Figure 2.7). The small amount of elements such as rays, longitudinal parenchyma and resin canals are important to tree functions and as aides to species identification of wood, but are of little significance to mechanical properties. Resin canals are not tissues like rays and longitudinal parenchyma, but tubular spaces between epithelium cells that contain resin in the sapwood. They appear in species such as pine, spruce, larch and Douglas-fir.

The structure of hardwoods is relatively complex, as illustrated in Figure 2.8. Hardwoods always contain at least four major types of cells, each of which may constitute 15% or more of the volume of a xylem. Lack of radial alignment in the cross-section, for most hardwood species, means hardwoods tend to be less transversely anisotropic

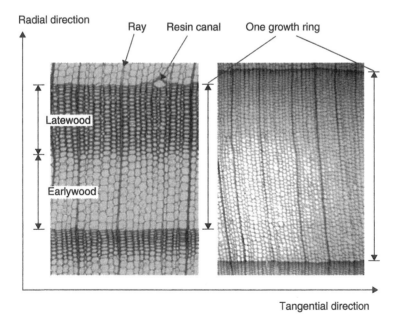

Figure 2.7 Transition in softwood from earlywood to latewood: abrupt in western larch, *Larix occidentalis* (left) and gradual in balsam fir, *Abies balsamea* (right).

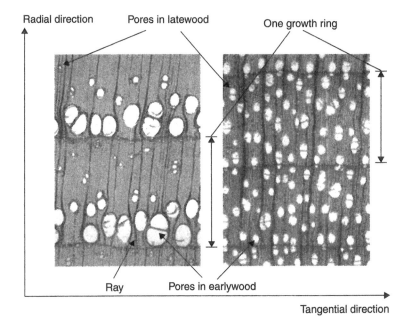

Figure 2.8   Distribution of pores in hardwoods: ring-porous wood (white ash, *Fraxinus amer-icana*, left) and diffuse-porous wood (yellow birch, *Betula alleghaniensis*, right).

than softwoods. Fibres occupy 50% or more of the volume and strongly influence mechanical properties. Vessels appear in the form of tissues, units of which are called vessel elements. Vessels have a transportation function and take up about 20% of the volume. The cross-section of a vessel element is known as a pore. Mechanical properties degrade with increase in the percentage of vessels by volume. Ray parenchyma account for 17% or more of the volume, and are responsible for different properties of hardwood in radial and tangential directions. However as already implied, any differences are normally not as pronounced as for softwoods. Longitudinal parenchyma range from 1–24% by volume, play a storage role in the growth of a tree, and have a slight negative effect on the mechanical properties for most species.

Clearly, it is necessary to at least group species into broad types of wood in any investigation related to properties of wood. Even then, findings need to be generalised with caution. There is no escaping the conclusion that microstructure has to be considered if one hopes to understand properties and behavioural mechanisms of wood.

2.1.6   Macrostructure

Macrostructural features of tree stems are essentially arranged longitudinally (parallel to the pith) and radially (relative to the pith) and can be understood via consideration of the cross-section, and radial and tangential sections (Figure 2.9). Each of these sections exhibits structural characteristics that suggest wood's inherent anisotropy.

As previously mentioned, the four parts in the cross-section are pith, xylem (heartwood and sapwood), cambium and bark. Figures 2.3 and 2.9 shows series of concentric growth rings (annual growth increments). A growth ring usually has relatively light

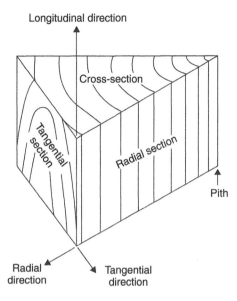

Figure 2.9 Definition of sections and axis directions in wood.

coloured large celled earlywood that is produced at the beginning of a growing season, and darker small celled latewood produced during the latter part of a growing season. Except for tropical trees, the cambium produces only one growth ring per year, which is why such growth increments are often called annual rings. The transition from earlywood to latewood differs between species as illustrated in Figures 2.7 and 2.8. In softwoods the transition may be abrupt exhibiting a distinct dense and dark latewood band, or gradual without a distinct latewood band (Figure 2.7). Latewood percentage defines latewood width as a proportion of growth ring width. This directly reflects the average density and determines many important end-use quality traits. In hardwoods transition between earlywood and latewood reflects the arrangement of pores. Figure 2.8 shows that white ash produces large pores in the beginning of a growing season, with the bulk of the fibres produced later in the growth season. Such trees are said to be ring-porous. By contrast, yellow birch produces pores having similar dimension over the whole growing season. Such species are referred to as diffuse-porous.

2.1.7 Variation of wood

Neighbouring locations within a tree may be similar, but they are never identical. Variation within individual trees can be as great as variation between trees of the same species, or variation between species. Within-tree variation is influenced by ageing of cambium, genetics and environmental conditions. As mentioned in the previous subsection, dimensions of cells vary across a growth ring. In conifers, for example, tracheids in the early part of a growth ring are relatively large with thin walls. This facilitates transportation of water and nutrients to the branches in the spring. Tracheids in the latter part of a growth ring are comparatively small with thick walls. This facilitates the development of strength and stiffness in stems and branches. The cambium regulates the quality and quantity of new wood in response to the prevailing stresses at any particular point. It has been said that the cambium optimises 'the mechanical design of the tree' (Mattheck and Breloer, 1994).

Figure 2.10 shows within-ring density profiles for three fast-grown species, where ring widths have been normalised to facilitate comparisons. The profiles shown are based on data of Jozsa (1996). Radial variations in density are frequently characterised at breast height and observed to follow one of three trends (Panshin and de Zeeuw, 1980) (Figure 2.11). Type I variation occurs in softwood species such as hard pines,

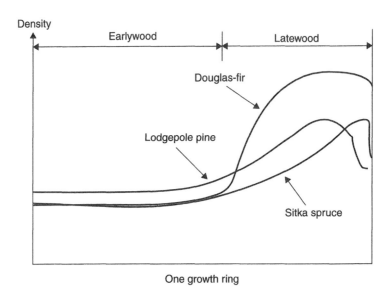

Figure 2.10   Schematic density profiles within one growth ring at breast height.

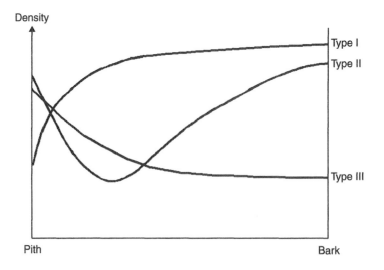

Figure 2.11   Diagrammatic radial variation of density.

Douglas fir and larch, and hardwood species such as mahogany, sweet gum, black poplar and eucalypt. Type II occurs in softwood species such as grand fir, black spruce, western red cedar and hemlock, and hardwood species such as black cottonwood. Type III variation occurs in softwood species such as white cedar, and hardwood species such as sugar maple, European beech and red oak.

Most often, the density of wood in the stem decreases moving from the base to the top, because the proportion of juvenile wood increases in that direction. This pattern exists in softwood species such as hard pines, Douglas fir and larch. For other conifers (e.g. Sitka spruce) the pattern is reversed. For hardwoods, the base-to-top variation follows either pattern and is usually small.

2.2   Pre-Service Damage in Wood

Wood components have some (pre-existing) damage, even if they have never been used. This can be in different forms and is variously referred to as pre-service damage, inherent defects or inherent damage. Damage may occur in standing trees, during harvesting and processing of stems, or as a result of biological agents.

2.2.1   Damage in standing trees

Damage in standing trees may be either gross damage, such as shakes or compression failure, or dispersed micro-cracks. It reflects inability of trees to counteract adverse stress situations due to the combination of growth stress, stress due to self-weight of supported parts of the tree system and stress due to environmental loads. Effects of wind, snow and ice loads or accidental impacts are often the metaphorical 'straw that broke the camels back'.

Shakes are gross longitudinal separations of wood in living trees. If the shakes are radial and pass through the pith they are known as pith or heart shakes. They can appear in single or multiple form. If the shakes occur between the growth rings partially or wholly encircling the pith they are known as ring shakes. Shakes can propagate under external forces in service and during seasoning. Compression failure is a result of permanent structural changes (kinks) in the cell walls, which are created by excessive compressive stress in the inner core of stems, especially in large stems. Eucalypts can contain severe compression failure, resulting in a marked decrease in the residual tensile and bending strength (Panshin and de Zeeuw, 1980). Frost cracks are another gross damage associated with non-tropical trees. Frost cracks are radial separations of cells in the wood and bark near the base of a tree, and most often occur in hardwoods, especially in old trees. Gross damages are usually detectable with the naked eye, and use of material that contains them can be avoided quite easily.

The Axiom of Uniform Stress espoused by Mattheck and Breloer (1994) states that '... a tree is a self-optimising mechanical structure. Its design therefore follows the rule for all such structures which, by definition, make as economic a use of their material as possible and are as strong as necessary. If such a structure is evenly loaded and all points on the surface have to withstand the same stress, it will have no overloaded areas (breaking points) and no under-loaded areas (wasted material). An optimal tree structure has uniform stress over the whole of its surface'. Under the Axiom of Uniform Stress a tree tends towards a shape that evens out surface stresses at each stage of growth. Moving radially from pith to bark, all locations in a stem or branch will have experiences comparable peak ratios of applied to critical stress, but not necessary at the same time in the life of the tree. As already mentioned, peak stresses are generated by a combination of growth and dead weight stresses, and stresses that are the consequence of environmental loads or mechanical impacts. Trees must balance minimisation of the wood content per tree with need for an adequate rate of survival of their population against life ending events like overturning or stem fracture. Most trees have just enough material in them to survive under local site conditions. It follows that localised (dispersed micro) damages that do not compromise the biological functioning of a tree are permissible and are not guarded against. Logic suggests that micro-damage at the microscopic (cell) or lower levels will always be present to some extent. Micro-damages can be due to either normal or shear stresses generated in a standing tree. It follows from the Axiom of Uniform Stress that damage in healthy trees that survive until harvest will be more or less equally distributed through their volume. Well dispersed pre-existing micro-damage has been observed via Scanning Elector Microscopy during real-time opening mode fracture tests on spruce (Vasic, 2000). In that instance damage appeared as fine micro-cracks in cell walls (<100 μm in length) that opened and closed elastically adjacent to, but not joining main cracks.

2.2.2 Damage due to harvesting and processing

Shakes in logs may develop parallel to grain when stems are felled. They can also develop immediately upon crosscutting a stem as a release of growth stresses. Other damage might occur during harvesting, transportation and handling of stems, but is not likely to be detectable with the naked eye. Further damage may appear in the course of sawing, machining and post-processing handling.

Drying is often the primary cause of processing induced damage in wood. Moisture movement is a time dependent process with the phase and rate of transport of fluids dependent on temperature within the wood and surface conditions (moisture carrying capability and flow rate of surrounding air). Because of its anisotropic nature, variability in its fluid content and moisture gradients at the time of drying, wood develops drying stresses and hence damage. The extent of drying damage is highly dependent on the drying method. Visible seasoning/drying damage is in the form of checks and splits that are separations of the wood cells in the grain direction.[3] The former does not extend through the thickness of lumber, but the latter goes from face to face. Deep checks and splits may considerably decrease the strength of lumber products, particularly in shear parallel to grain and tension perpendicular to grain. During kiln drying, the last step is conditioning, the purpose of which is to create compressive surface stress perpendicular to grain. This closes many splits and checks, and so what is visible is not their true extent. Fine scale drying damage is undetectable in practical situations but always occurs to some extent unless drying is exceptionally slow (over many years). As pointed out by Zhou and Smith (1991), presence of grain distortions means

---

[3] Checks and splits reflect wood's inherent weakness under tension perpendicular to grain, which in turn reflects that there are few cells aligned radially or tangentially compared to parallel to grain.

that low quality lumber products will experience significant damage whatever the type of drying.

Improper sawing and machining may generate strength-reducing features such as diagonal grain, breakage of knots and knotholes. Diagonal grain is the result of sawing parallel to the pith of a stem with pronounced taper or crookedness. The most severe situation is combination of spiral grain and diagonal grain, which causes so-called cross grain. Diagonal or cross grain has a serious detrimental effect on mechanical properties parallel to grain. Broken knots and knotholes are stress raisers causing localised weakness and decreasing most mechanical properties.

2.2.3 *Damage due to biological agents*

Damage in wood can occur from attack by bacteria, fungi, insect and/or marine borers in the living tree or after the tree has been felled, transported, processed and in end-use.

Bacteria cause damage such as wet pockets in trees. Wet pockets (wetwood) are anomalous zones with high moisture content, normally occurring in heartwood close to the boundary with sapwood and often penetrating to the pith. They occur in hardwood species such as oak, poplar, maple and birch, and softwood species such as true fir, eastern white pine and hemlock. Common locations are the base area in association with damaged roots and higher positions associated with branch knots or wounds. Shakes and frost cracks are often concurrent with wet pockets, with the latter being the sites for initiation of the mechanical damage.

Fungi grow in wood causing decay and mould, and eventually measurable change in physical and mechanical properties. The extent of fungal effects is highly dependent on wood species, type of fungus, and intensity and length of an attack. Brown rots and white rots are very common decay fungi. Brown rots aggressively attack cellulose and hemicellulose in cell walls, but not lignin. White rots attack all major organic components. Typically, white rots degrade lignin more rapidly than cellulose and hemicellulose. Even in the very early stages of development, attack from either brown or white rot fungi significantly effects mechanical properties of wood. Moulds on the other hand do not have significant effect on mechanical properties of wood, and so these are often ignored from a mechanics perspective.

Insects like termites and beetles can cause enormous damage to wood both in the forest and in end-use situations. The natural geographic range of various insects is restricted due to climate, but global warming and trade in forest products are bringing about some fairly rapid change in dispersions. For example, termites are now not uncommon in some milder regions of Canada. Marine borers inhabit the coastal waters in many regions of the world. They can attack aggressively and destroy logs rapidly during their transportation and storage in waters.

2.3 Physical and Mechanical Properties of Wood

Properties of wood by necessity reflect its structure. This section discusses key physical and mechanical properties and factors influencing them, as a precursor to other chapters.

2.3.1 Density and porosity

Density ρ is an important physical index, as it correlates with most mechanical and some other physical properties. It is defined as:

$$\rho = \frac{M}{V} \tag{2.1}$$

where M is the mass (kg) and V the volume (m^3) of the piece of wood used for its determination. Since moisture changes both the mass and the volume (due to shrinkage or swelling of cell walls), density is moisture dependent. Conditions for its assessment must be specified. In scientific circles, basic density ρ_b is the preferred measurement, since both the mass and volume can be defined unambiguously:

$$\rho_b = \frac{oven\text{-}dry\ mass\ of\ wood}{volume\ of\ green\ (saturated)\ wood} \tag{2.2}$$

The term relative density RD (or specific gravity SG) is used in much of the wood science literature. It is defined as:

$$RD = \frac{oven\text{-}dry\ mass\ of\ wood}{mass\ of\ a\ displaced\ volume\ of\ water} \tag{2.3}$$

When determining the mass of a displaced volume of water, the moisture content has to be defined. Figure 2.12 illustrates the arithmetic distribution of RD for 65 hardwood

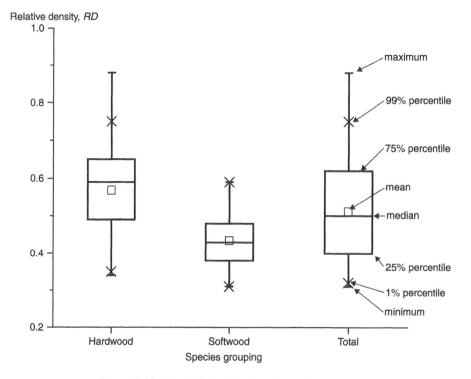

Figure 2.12 Typical distributions of relative density.

and 47 softwood species grown in the USA, with *RD* based on oven-dry mass and volume at 12% moisture content. Although not representative of all wood, this indicates the much greater range of *RD* in hardwoods than in softwoods. In general, both heaviest and lightest species are hardwoods. For examples, African blackwood (*Dalbergia melanoxylon*) has a *RD* as large as 1.12, but Balsa (*Ochroma pyramidale*) only 0.12. Relative density may be classified into three groups: light (*RD* < 0.36), medium (0.36 ≤ *RD* < 0.50), and heavy (*RD* ≥ 0.50) (Schiewind, 1989).

Void spaces in wood includes cell lumen, pit openings,[4] pit cavities, and intercellular spaces. Once these spaces are filled with water, all wood will sink. Remembering that wood substance has an average density of 1500 kg/m^3 (see Section 2.1.3) the density of wood and its porosity *P* are related:

$$P = \left(1 - \frac{RD_{oven\text{-}dry}}{1.50}\right) \times 100(\%) \qquad (2.4)$$

For instance, African blackwood has a *P* of about 25% but balsa about 92%. Porosity gives an estimate of the maximum amount of water that can be held in a piece of wood, and determines the density of wood.

2.3.2 Moisture content

Water is an essential component of a living tree, and its weight can exceed that of the wood substance. The total amount of water in a given piece of wood is usually expressed by the moisture content, *m*:

$$m = \frac{mass\ of\ water\ in\ wood}{mass\ of\ oven\text{-}dry\ wood} \times 100(\%) \qquad (2.5)$$

Most commercially important woods have an *m* of about 60% for the heavy wood and 200% for the light wood at the time of felling. Generally, softwoods have a larger *m* value in sapwood than in heartwood. Hardwoods do not have such a fixed mode and vary from species to species. Water in wood exists as free water and bound water. Free water exists in cell lumen and cavities in liquid and/or vapour form. Bound water is held within cell walls by hydrogen bonds and van der Waals forces. The removal of bound water requires much more energy than the removal of free water. When wood dries the liquid water in cell lumens and cavities leaves first. The condition at which only bound water remains is termed the fibre saturation point, *FSP*. Only bound water influences mechanical and some physical properties of wood. Below the *FSP*, many properties of wood change. The *FSP* is expressed as a moisture content value, which varies among species but is usually within the range 25–30%.

Due to the existence of hydrogen bonding sites in hydroxyl groups present in non-crystalline cellulose, hemicellulose and lignin wood is a hygroscopic material. In the saturated condition 'attracting' sites have five or six layers of attached water molecules. Dried wood has only one layer of water molecules at an *m* of about 6%. Wood constantly exchanges moisture with its surrounding atmosphere even in an equilibrium

---

[4] Pits are gaps in the secondary wall but the primary wall remains in place and serves as a membrane that controls the movement of liquids and vapour.

system. This depends upon the surrounding atmosphere's moisture carrying capability that is proportional to its temperature, relative humidity and the rate of flow. For any given surrounding atmosphere, there is a corresponding equilibrium moisture content *EMC* at which inward adsorption equals outward desorption (Figure 2.13). For example, at a temperature of 20°C and relatively humidity of 65%, the *EMC* of wood has a value of about 12% for most species. *EMC* decreases with any increase of temperature at a fixed relative humidity.[5]

At a given environmental condition, the *EMC* reached during an adsorption process is always a little lower than that during a desorping process, Figure 2.13. This phenomenon is called sorption hysteresis, with the difference between desorption and adsorption equilibrium states Δm being essentially independent of species. The value of Δm varies from about 0.2% for short and thin wood pieces to 2.5% for long and thick wood pieces. The ratio of adsorption *EMC* to desorption *EMC* at room temperature is reported to be between 0.8 and 0.9 (Hoffmeyer, 1995). The sorption hysteresis leads to reduced dimensional change in dried wood exposed to cyclic humidity. Engineered wood products have lower *EMC* values than wood, due to the introduction of adhesive, heat and/or chemical treatment of wood during their manufacture.

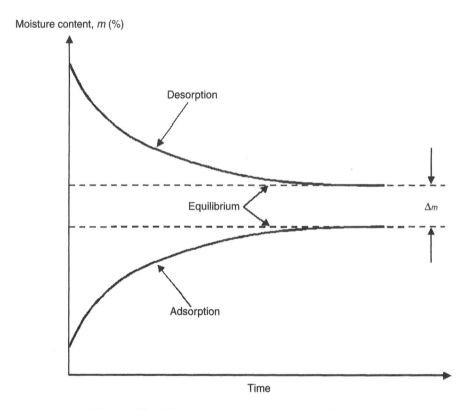

Figure 2.13 Adsorption and desorption curves for wood.

---

[5] *EMC* also depends upon, and increases with, any increase in the rate of flow of surrounding air. Quoted *EMC* values presume still air.

2.3.3   *Dimensional changes in wood*

Dimensional changes in wood are primarily the result of change in the bound water content of cell walls. Wood begins to shrink as it dries below the *FSP*, and continues to shrink until $m = 0$. Conversely, wood swell as it absorbs moisture until $m = FSP$. No further swelling occurs if $m > FSP$. Dimensional change is usually expressed as a percentage of the original value:

$$Shrinkage = \frac{decrease\ in\ dimension}{green\ (saturated)\ dimension} \times 100(\%) \qquad (2.6)$$

$$Swelling = \frac{increase\ in\ dimension}{oven\text{-}dry\ dimension} \times 100(\%) \qquad (2.7)$$

Total shrinkage (from green to oven-dry condition) of 'normal' wood is about 0.1–0.3% in the longitudinal direction, 3–5% in the radial direction, and 6–10% in the tangential direction. This reflects directionally dependent arrangements of cells. Differential shrinkage is the ratio of the tangential shrinkage to the radial shrinkage and on average is about 2 for most species, with values for softwoods normally larger than for hardwoods due to the radial alignment of cells in softwoods (see Section 2.1.5). Wood of high density has relatively large shrinkage because of the low porosity (proportion occupied by cell walls is high). The longitudinal shrinkage of reaction wood and juvenile wood is larger than for normal wood because the former two have greater microfibril angles.

Dimensional changes in wood are assumed linear from oven-dry to green conditions (USDA, 1999), with shrinkage estimated according to the equation:

$$S_{MC} = S_O \left[\frac{FSP - m}{FSP} \right] \qquad (2.8)$$

where S_{MC} is the shrinkage from the green (*FSP*) condition to the m of interest, and S_O is the total shrinkage from the green to oven-dry conditions. The total volumetric shrinkage, S_V, can be employed to predict the density at a given moisture content $\rho_m (0 \leq m \leq FSP)$:

$$\rho_m = \frac{\rho_b}{1 - S_V} \qquad (2.9)$$

2.3.4   *Mechanical properties of wood*

For engineering uses, interest in wood has been strongly focussed on behaviour when stressed parallel to grain, because that is its strong orientation with respect to strength and stiffness. Figure 2.14 illustrates shapes of stress-strain curves for wood loaded in tension and compression parallel to grain. Curves contain linear, nonlinear and failure regions. Any change in dimension or shape of a specimen is not fully recoverable in the nonlinear region after removal of stress and permanent set occurs. Beyond the peak stress in tension wood is brittle and fails at small strain. In compression it is apparently ductile and retains ability to carry stress until very large strain. For normal wood, the peak stress is much higher in tension than compression, which is attributed

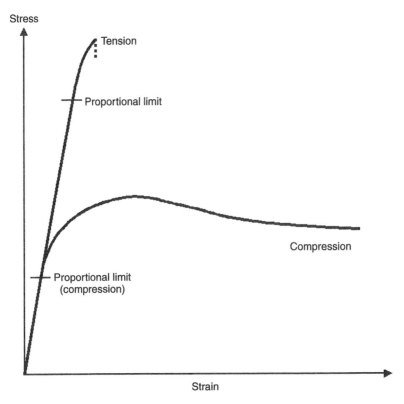

Figure 2.14   Stress versus strain curves for wood loaded in tension and compression parallel to grain.

to covalent bonding within cellulose and the physical structure and arrangement of cells. For stress perpendicular to grain the response is essentially brittle in tension, but the stress-strain response in compression is complex. In an approximate sense, compression responses quoted in the literature give the appearance of an elasto-plastic response with or without hardening as might be observed for metals. However, what is observed depends to a large extent on how the test is made.[6] As discussed in Chapter 3, this should not in any way be construed as implying there actually is plastic strain. Perpendicular to grain the peak stress in tension is often less than in compression, because few cells are provided that can resist tension, especially if stress is applied in the tangential direction (see Section 2.1.5). In shear the response is quasi-brittle with quite low strength.

The indexes of mechanical properties of wood most commonly used in wood science/technology literature are illustrated in Table 2.3. Figure 2.15 show illustrative distributions of two key mechanical properties: modulus of rupture (apparent bending strength), and modulus of elasticity parallel to grain. The data are based on USDA

---

[6] In the wood science/technology literature, quoted mechanical properties were usually determined based on standardised test methods that do not necessarily yield true properties (from a mechanics perspective). The primary purpose of data is as comparative measures (indexes) of performance of different species.

Table 2.3   Common mechanical properties of clear wood

Loading Type	Illustration	Mechanical index
Compression parallel to grain		Maximum compressive strength, modulus of elasticity, proportional limit
Compression perpendicular to grain		Proportional limit strength
Shear parallel to grain		Shear strength
Static bending		Modulus of rupture, modulus of elasticity, proportional limit
Tension parallel to grain		Tensile strength, modulus of elasticity, proportional limit
Tension perpendicular to grain		Tensile strength

(1999) that covers 65 hardwood and 47 softwood species produced in the USA. What is immediately apparent is the strong correspondence between the relative ranges of mechanical properties and relative density (Figure 2.12). Indeed, to a reasonable approximation, all mechanical properties of clear wood can be taken as linearly proportional to density. Figure 2.16 shows that both modulus of elasticity and modulus of rupture have a linear relationship to density.

Above discussion relates to behaviour of normal wood without imperfections such as knots and grain deviation, so-called clear wood.

2.3.5   Key factors affecting mechanical properties of wood

Other than density, mechanical properties of wood are strongly influenced by its natural characteristics (e.g. knots and grain deviation) and environmental service conditions (e.g. moisture content and temperature).

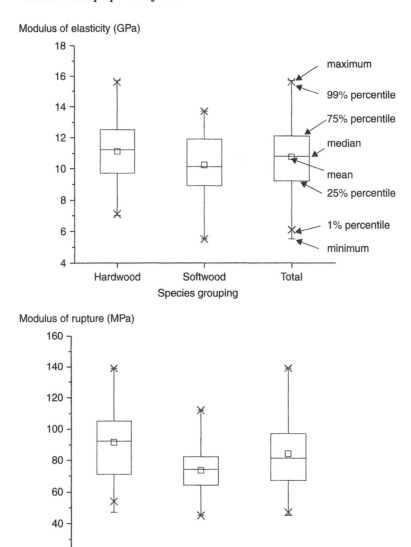

Figure 2.15   Approximate distributions of modulus of elasticity and modulus of rupture.

Effect of grain deviation

General grain deviations (general slope of grain) are the result of natural spiral grain and deviations introduced during processing of components. The general slope of grain is observed away from localised perturbations existing around features such as knots (local slope of grain). In lumber, when the grain direction is parallel to the edges the lumber is said to have straight grain. If however, the grain direction is not parallel to the edges the grain is 'deviated'. Both general and local slope of grain conditions affect

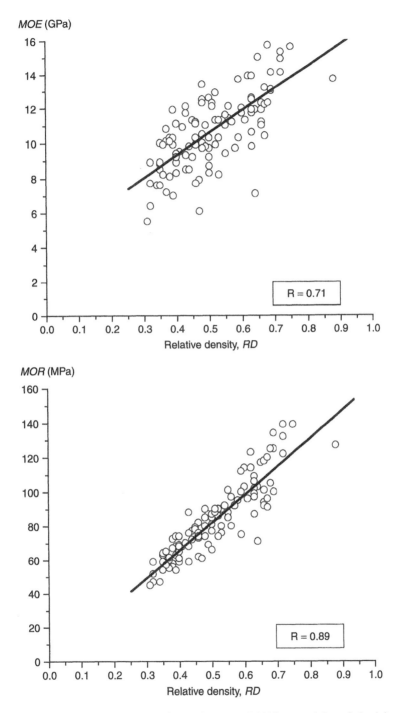

Figure 2.16 Relationship between modulus of rupture (*MOR*) or modulus of elasticity (*MOE*) and relative density (*RD*).

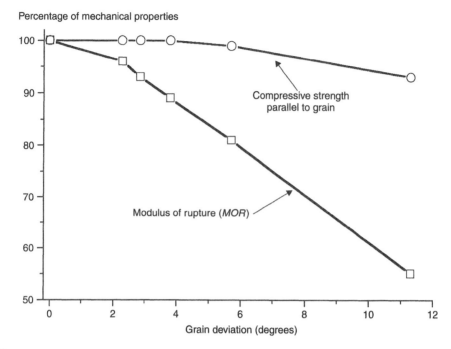

Figure 2.17 Change in mechanical properties of wood members as a function of grain deviation.

the apparent mechanical properties of lumber components. Figure 2.17 illustrates the effect of grain deviation on modulus of rupture and compressive strength in the parallel to grain direction (UDSA 1999). A grain deviation of 2.9 degrees causes 7% loss in modulus of rupture. The effect of grain deviation on modulus of elasticity parallel to grain is illustrated in Figure 2.18, based on data of Gong (1990). It can be seen that there is rapid decrease in the property as the slope increases from 0 to 45 degrees, but thereafter the trend levels off. It is reported that a grain deviation of 15 degrees can result in 45%, 70% and 80% reductions in tensile strength, modulus of rupture and compressive strength respectively, relative to parallel to grain values (Desch and Dinwoodie, 1981).

Effect of knots

Knots have negative effects on most mechanical properties of wood because they distort the flow of the grain. Consequently eccentricities inevitably develop in the flow of forces within components containing knots. Whatever the nominal stress condition for a lumber component there will be stress components perpendicular to grain and shear stress parallel to grain. As mentioned in Section 2.3.4, these are not stress conditions under which wood's strength is effectively utilised. How critical this is depends on the positioning of the knot(s) within a component, its size, soundness, and geometry (Figure 2.19). Reductions in properties are generally proportional to size of a knot, and edges knots usually have more serious effect on mechanical properties than centre-line knots. Encased knots usually create less localised grain deviation than intergrown

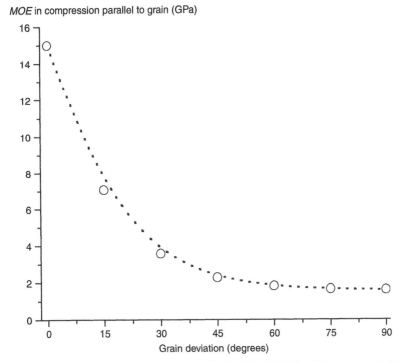

Figure 2.18   Grain deviation versus modulus of elasticity (*MOE*) of Japanese ash (*Fraxinus mandshurica*) in compression parallel to grain.

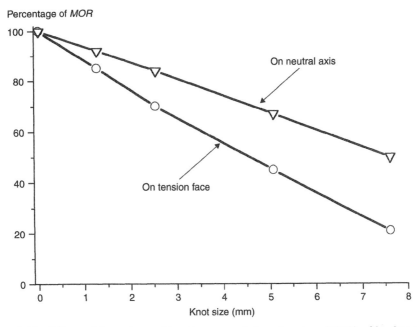

Figure 2.19   Effects of knot size and location on modulus of rupture (*MOR*) of lumber with a cross-section of 38 × 140 mm.

ones, and so the resulting degradation in strength is less. Knots have most effect on tensile strength and modulus of rupture parallel to grain, and only slight effect on the perpendicular to grain compressive strength, and modulus of elasticity and shear strength parallel to grain.

Effect of moisture content

Most mechanical properties of defect free (clear) wood increase with decreasing moisture content below the *FSP*. This is because many new hydrogen bonds are generated in the microfibrils with removal of water molecules from cell walls, resulting in an increase of the crystalline regions. The overall relationship between mechanical properties and moisture content is not linear, but it is approximately so from 8–22% moisture content. The change in selected mechanical properties due to unit change in m is given in Table 2.4. As the table shows, compression properties are most sensitive to moisture change. Figure 2.20 shows the influence of moisture content on modulus of elasticity in tension parallel to grain, based on data of Gong (1990). It can be seen that there is linear increase in modulus of elasticity for $10 < m < 24\%$, but a drop occurs for $m < 10\%$. As mentioned in Section 2.3.2, at m of 6%, there only exists one layer of water molecules that is directly attached to cell walls by the hydrogen bonding. Change in this layer of water molecules may degrade the mechanical properties of wood. Although the influence of moisture on mechanical properties is strong for many wood products, its effect on strength properties of low quality lumber is quite weak (Madsen, 1975; Zhou and Smith, 1991; Hoffmeyer, 1995), but that is not the case for stiffness properties.

Effect of temperature

The mechanical properties of wood usually decrease with increasing temperature, and increase with decreasing temperature. Under normal end-use conditions the effect of temperature is most evident at cold temperatures, because freezing solidifies the moisture within the wood. However, any gain is lost upon thawing. Freezing can cause

Table 2.4 Average changes in mechanical properties of clear wood due to one percent change in moisture content (based on data of Bodig and Jayne (1982) and Hoffmeyer (1995))

Property	Change (%)
Compressive strength parallel	5
Compressive strength perpendicular	5.5
Shear strength parallel	3
Modulus of rupture parallel	4
Modulus of elasticity parallel	2
Tension strength parallel	2.5
Tension strength perpendicular	1.5

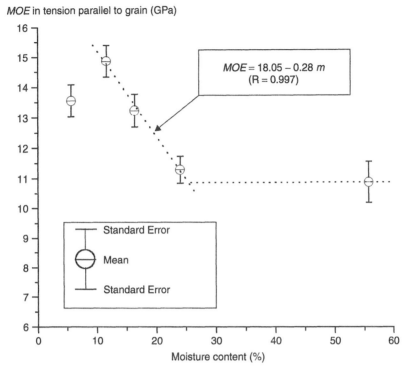

MOE in tension parallel to grain (GPa)

$MOE = 18.05 - 0.28\ m$
(R = 0.997)

Standard Error

Mean

Standard Error

Moisture content (%)

Figure 2.20   Effect of moisture content on modulus of elasticity (*MOE*) of Japanese ash (*Fraxinus mandshurica*) in tension parallel to grain.

damage as the water in voids expands. Modulus of elasticity and compression strength perpendicular to grain are the most temperature-sensitive properties. However, there is little permanent reduction in mechanical properties of wood when temperature does not exceed 100°C. As a rule of thumb, an increase in temperature of 1°C produces 1% reduction in most mechanical properties between −20 and +65°C. As mentioned in Section 2.1.3, lignin begins to soften when the temperature reaches +170°C, suggesting that wood will begin to lose strength. Severity of temperature effects depends upon the moisture content, with its effect being considerably greater at higher moisture contents.

Effect of other factors

Many factors can influence mechanical properties of wood, such as annual ring orientation, latewood percentage, reaction wood, juvenile wood, microscopic features (microfibril angle, fibre length, cell shape and dimension, and cell wall thickness), duration of load and exposure to chemical and biological agents. A number of these are discussed in later chapters in the context of how they influence fracture and fatigue behaviour of wood and wood-based products. For those readers who wish to gain a deeper understanding of how trees grow, the nature of wood and how wood science people think about the topic, it is suggested they start with texts by Kollmann and Côté (1968), Panshin and de Zeeuw (1980), Desch and Dinwoodie (1981) and Haygreen and Bowyer (1996).

2.4   References

Alhasani, M.A. (1999) *Growth stresses in Norway spruce*, Report TVBK-1016, Division of Structural Engineering, Lund Institute of Technology, Lund, Sweden.

Bamber, R.K. (1978) 'Origin of growth stresses', *Forpride Digest*, **8**(1): 75–96.

Bodig, J. and Jayne, B.A. (1982) Mechanics of Wood and Wood Composites, Van Nostrand Reinhold, New York, NY, USA.

Desch, H.E. and Dinwoodie, J.M. (1981) Timber: Its structure, properties and utilisation, Macmillan, UK.

Dinwoodie, J.M. (1989) Wood: Nature's cellular, polymeric fibre-composite, The Institute of Metals, London, UK.

Gong, M. (1990) 'A study of wood elasticity', MScE thesis, Nanjing Forestry University, Nanjing, Jiangsu, China.

Haygreen, J.G. and Bowyer, J.L. (1996) Forest Products and Wood Science: An introduction, Iowa State University Press, Ames, IW, USA.

Hoffmeyer, P. (1995) 'Wood as a building material', Lecture A4. In: Timber Engineering Step 1: Basis of design, material properties, structural components and joints, Ed. by Blass, H.J., Aune, P., Choo, B.S., Gorlacher, R., Griffith, D.R., Hilson, B.O., Racher, P. and Steck, G., Centrum Hout, Almere, The Netherlands: A4/1–A4/21.

Jozsa, L.A. (1996) 'Timber management toward wood quality and end-value: stemwood density trends in second-growth British Columbia softwoods', *Proceedings of CTIA/IUFRO International Wood Quality Workshop*, Quebec City, QC, Canada.

Kollmann, F.F.P. and Côté, W.A. (1968) Principles of Wood Science and Technology I: Solid wood, Springer-Verlag, New York, NY, USA.

Madsen, B. (1975) *Moisture content-strength relationship for lumber subjected to bending*, Structural Research Series Report No. 11, Dept. Struc. Eng., University of British Columbia, Vancouver, BC, Canada.

Mattheck, C. and Breloer, H. (1994) The body language of trees: A handbook for failure analysis, Research for Amenity Trees No. 4, The Stationary Office, London, UK.

Munch, E. (1938) 'Statics and dynamics of the cell wall's spiral structure, especially in compression wood and tension wood', *Flora*, **32**: 357–424.

Panshin, A.J. and de Zeeuw, C. (1980) Textbook of Wood Technology, McGraw-Hill, New York, NY, USA.

Schniewind, A.P. (1989) Concise Encyclopedia of Wood and Wood-based Materials, Pergamon Press, Oxford, UK.

USDA (1999) Wood Handbook: Wood as an engineering material, Forest Products Laboratory, Forest Service, United States Department of Agriculture, US Government Printing Office, Washington, DC, USA.

Vasic, S. (2000) 'Applications of fracture mechanics to wood', PhD Thesis, University of New Brunswick, Fredericton, NB, Canada.

Zhou, H. and Smith, I. (1991) 'Influences of drying treatments on bending properties of plantation–grown white spruce', *Forest Products Journal*, **41**(2): 8–14.

Appendix: Notation

EMC = equilibrium moisture content (%)
FSP = fibre saturation point (%)
m = moisture content (%)
M = mass (kg)

MFA = microfibril angle
MOE = modulus of elasticity (GPa)
MOR = modulus of rupture = apparent bending strength (MPa)
P = porosity (%)
R = coefficient of determination
RD = relative density
SG = specific gravity
S_{MC} = shrinkage from the green (*FSP*) condition to m of interest
S_O = total shrinkage from the green to oven-dry conditions
S_V = total volumetric shrinkage
V = volume (m^3)
ρ = density (kg/m^3)
ρ_b = basic density (kg/m^3)
ρ_m = density a given $m, 0 \leq m \leq FSP$ (kg/m^3)

3

Mechanical Behaviour of Wood: Concepts and Modelling

3.1 Material Complexity and Modelling Levels

Wood is by its nature heterogeneous with structured discontinuity, even though both these traits are commonly ignored. Its nature and properties can be conceptualised and modelled on many levels, ranging from chemical constituents to gross behaviour of massive wood (Figure 3.1). Engineers usually make arbitrary assumptions regarding the nature of wood and wood-derived materials for analytical convenience. It is commonly assumed that wood is an elastic, solid material with well-defined directional dependence of its physical structure and properties. However it is only under certain situations that wood is solid, elastic, or has directionally well defined properties. As discussed in Sections 3.2, 3.4.1, 3.4.5 and 3.5.1, presence of hygroscopic amorphous regions in cell walls means that physical and mechanical properties of wood and derived materials are highly sensitive to the present and or past moisture-states of structural components. Wood exhibits glassy (solid) and rubbery states as a function of its temperature. Glass transition temperature T_g of a material is the temperature at which there is abrupt change in stiffness, and it mediates between glassy and rubbery regimes of polymers. The apparent rubbery transition of wood reflects the T_g values of hemicellulose and lignin that are about 80°C and 90°C respectively (Hillis and Rozsa, 1985). There is significant weight loss when the surface of wood is heated to about 250°C due to evolution of bound water. Between 250°C and 400°C active pyrolysis takes place with wood decomposing into flammable gases, tars and char residue (MacKay, 1967). Both physical and mechanical properties of wood are nonlinear with regard to the temporal history. The amount of moisture shrinkage or swelling is a function of the moisture history (Sections 2.3.2 and 2.3.3). Accumulated strain is a function of moisture, thermal and stress histories and their non-linear interaction. Clearly, an analyst must possess more than rudimentary knowledge of wood if he/she is to successfully predict deformation or failure processes.

Fracture and Fatigue in Wood   I. Smith, E. Landis and M. Gong
© 2003 John Wiley & Sons, Ltd   ISBN: 0-471-48708-2 (HB)

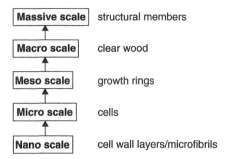

Figure 3.1 Levels at which mechanical properties of wood are modelled.

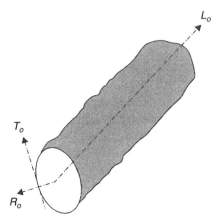

Figure 3.2 Wood in a stem represented as cylindrically orthotropic material(L_o = longitudinal or grain direction, R_o = radial direction, T_o = tangential direction).

In analysis wood is often taken to be cylindrically orthotropic material (Figure 3.2). Although this neglects effects of local perturbation in wood structure around features such as knots, radial variation in the growth rings, and taper, crook and sweep in logs, it is sound underpinning to more detailed representations. At a finer levels of resolution properties vary in both longitudinal, L_o, and radial, R_o, directions (Smith *et al.*, 1991; Kaya and Smith, 1993). Conversely, a simplification made for engineering design purposes is presumption of transverse isotropy, i.e. ignoring directional dependence of properties in the radial-tangential, R_o–T_o, plane. So-called mechanical, or indeed other, properties of wood are only constant under well-defined conditions. Ignoring this can lead to poor modelling and misapplication of the material.

At what level of detail mechanical properties of wood need to be represented depends on an analyst's purpose. For example, it is often adequate to consider wood as an elastic and transversely isotropic material in fracture analysis of massive members containing large holes or notches. However, if one needs to understand behaviour of small-scale fatigue specimens it is necessary to consider the microstructure. A monocular treatment of properties of wood does not suffice if one is to benefit from reading this book or understand fatigue and fracture of wood.

This chapter starts with models for representing mechanical properties at the micro scale, and eventually arrives at representation of wood on the massive scale. Particular emphasis is placed on the macro scale (clear wood), because this is the level of representation most often used in numerical fracture analysis. The aim is provision of background and support information to other chapters, rather than a comprehensive treatise on wood mechanics.

3.2   Micro Scale: Cell Level

Wood is a two-phase material with layers in cell walls that consist of crystalline cellulose forming fibres in a matrix of non-crystalline cellulose, hemicellulose and lignin, with minor amounts (5–10%) of extraneous material. Non-crystalline regions absorb water and it is this feature that leads to a moisture dependence of mechanical properties of wood. All layers of cell walls contain microfibrils, except for the middle lamella (Figure 2.5). Although the types of cells that appear and how they are arranged depends on the tree species, the arrangement is always complex (see Chapter 2). Notwithstanding the complexity, some success has been achieved in predicting stiffness properties of cell walls from properties of the chemical constituents. Focus is on the S_2 layer because of its dominant influence on the cell wall (Table 2.1). Thickness and relative orientation of microfibrils are the primary factors influencing contribution of a layer to wall stiffness in a co-ordinate direction.

Stiffness and shrinkage properties of cell wall layers and complete walls can be predicted from the corresponding responses of chemical constituents (Sakurada *et al.*, 1962; Mark, 1967; Cave, 1978; Persson, 1997). Ignoring small water absorbing non-crystalline regions, cellulose is approximately transversely isotropic and moisture insensitive, with the elastic stiffness matrix D_C (Cave, 1978; Persson, 1997):

$$D_C = \begin{bmatrix} 137 & 14 & 14 & 0 & 0 & 0 \\ 14 & 27 & 14 & 0 & 0 & 0 \\ 14 & 14 & 27 & 0 & 0 & 0 \\ 0 & 0 & 0 & 6.6 & 0 & 0 \\ 0 & 0 & 0 & 0 & 6.6 & 0 \\ 0 & 0 & 0 & 0 & 0 & 6.6 \end{bmatrix} \text{ GPa} \tag{3.1}$$

Hemicellulose, although also transversely isotropic, is moisture sensitive with the elastic stiffness matrix D_H (Cave, 1978; Persson, 1997):

$$D_H = \begin{bmatrix} 8 & 2 & 2 & 0 & 0 & 0 \\ 2 & 4 & 2 & 0 & 0 & 0 \\ 2 & 2 & 4 & 0 & 0 & 0 \\ 0 & 0 & 0 & 1 & 0 & 0 \\ 0 & 0 & 0 & 0 & 1 & 0 \\ 0 & 0 & 0 & 0 & 0 & 1 \end{bmatrix} c_H(m) \text{ GPa} \tag{3.2}$$

where $c_H(m) = h\frac{0.1+0.9(1-m)}{1+a}$ ($0.05 \leq m \leq FSP$), m is moisture content, h is an empirical 'stiffness' parameter with a typical value 6.0 (Cousins, 1976), a is an 'area compensation' factor relative to the dry state with a typical value 1.0, and *FSP* is the fibre saturation point. *FSP* usually lies between 24 and 30%, depending upon species.

Lignin is isotropic and moisture sensitive with the elastic stiffness matrix D_L (Cave, 1978; Persson, 1997):

$$D_L = \begin{bmatrix} 4 & 2 & 2 & 0 & 0 & 0 \\ 2 & 4 & 2 & 0 & 0 & 0 \\ 2 & 2 & 4 & 0 & 0 & 0 \\ 0 & 0 & 0 & 1 & 0 & 0 \\ 0 & 0 & 0 & 0 & 1 & 0 \\ 0 & 0 & 0 & 0 & 0 & 1 \end{bmatrix} c_L(m) \quad \text{GPa} \tag{3.3}$$

where $c_L(m) = \frac{1+2(1-m)}{1+a}$ $(0.05 \le m \le FSP)$.

Stiffness matrices for cell walls can be developed by transforming stiffness matrices for wall layers so that they apply to the co-ordinate system for the cell, and weighting of stiffness terms in proportion to layer thickness. The transformation for the S_2 layer is between local Cartesian co-ordinate directions as defined by the microfibril alignment, and cell wall co-ordinate directions according to the relationship:

$$\hat{D}^{S_2} = G^T D^{S_2} G \tag{3.4}$$

where $\hat{D}^{S_2}$ is the stiffness matrix in cell wall co-ordinates, D^{S_2} the stiffness matrix in microfibril co-ordinates, and G the transformation matrix. Following the transformation process, the stiffness of the S_2 layer in the longitudinal cell axis direction becomes (Persson, 1997):

$$\hat{D}^{S_2}_{LL} = [D^{S_2}_{11} - 2D^{S_2}_{12} + D^{S_2}_{22} - 4D^{S_2}_{44}]\cos^4\varphi + [2D^{S_2}_{12} - 2D^{S_2}_{22} + 4D^{S_2}_{44}]\cos^2\varphi + D^{S_2}_{22} \tag{3.5}$$

'Local stiffness terms' are for cellulose as given in Equation (3.1), and φ is the microfibril angle for the S_2 layer. If the S_2 layer takes up a proportional cell wall volume p, and assuming the stiffness of the cellulose layer in the L direction is R' times the stiffness of other layers, the stiffness of the cell wall in the L direction is:

$$D^{cw}_{LL} = \left[p + \frac{(1-p)}{R'} \right] \hat{D}^{S_2}_{LL} \tag{3.6}$$

If, for example, $p = 0.8$ and $R' = 5$, $D^{cw}_{LL} = 0.84\hat{D}^{S_2}_{LL}$ (Persson, 1997).

In the three dimensions $(L - R - T)$ or two dimensions $(R - T$ plane) the structure of individual cells is usually modelled assuming regular honeycomb cross-section geometry (Price, 1928; Gibson and Ashby, 1988; Koponen *et al.*, 1989, 1991; Kahle and Woodhouse, 1994). Cells have often been taken as hexagon shaped in the $R - T$ plane, which means that well-known linear elastic stiffness analysis techniques, as employed in analysis of plane frames, can be used to estimate forces within walls under in-plane loads. This yields closed form expressions that require axial and flexural rigidities of cell walls as input. Wall rigidity terms are deduced through experiments, or from first principles commencing with chemical composition of wall layers. Closed form expressions have also been developed for in-plane elastic stability of cell walls due to in-plane loads (Gibson and Ashby, 1988). For individual cells, it is practical to apply numerical, finite element, techniques in two or three dimensions, starting with microfibrils as basic building blocks, and accounting for layered cell wall structure (Preston, 1974; Persson, 1997).

In themselves sub-cell and cell level models cannot produce output reflective of how wood behaves. They are mainly useful as building blocks for analysis of larger systems (higher structural scales).

3.3 Meso Scale: Growth Ring Level

Radial variation of apparent mechanical properties over a growth ring relates to variation in density. A schematic trend of variation in basic density (dry mass divided by green volume) within a spruce growth ring is shown in Figure 3.3. Cell wall material has a bulk density of about 1500 kg/m^3 (1.5 g/cm^3) for all living plants (see Section 2.1.3). There exists the general relationship that density is 1500 S^{cw} kg/m^3(1.5 S^{cw} g/cm^3), with S^{cw} being the cell wall ratio. Average density is:

$$\rho_{av} = 1500 S_{av}^{cw} \quad \text{kg/m}^3 \tag{3.7}$$

Persson (1997) suggests dividing the growth ring into earlywood, transition-wood and latewood zones of constant density (Figure 3.3). For spruce the width of the transition zone l_t is taken as a constant proportion of the ring width l_r ($l_t = 0.2 l_r$). The width of the latewood zone l_l is taken to be 0.2 mm. Thus, average density in the radial direction is:

$$\rho_{av-R} = \rho_e + 0.2(\rho_t - \rho_e) + (\rho_l - \rho_e)\frac{l_l}{l_r} \tag{3.8}$$

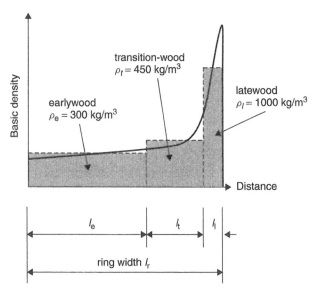

Figure 3.3 Variation in basic density over a growth ring in spruce (based on Persson, 1997).

There will be variations in widths of zones within growth rings even within a species depending on growing conditions and other factors, but similar strategies to that underpinning Equation (3.8) can be envisaged for all species of wood.

As an extension of Equation (3.6), the stiffness of wood within one or a number of adjacent growth rings in the longitudinal direction D_{LL} can be estimated based on the cell wall area ratio (Persson, 1997):

$$D_{LL} = \frac{\rho_{av}}{\rho_o} D_{LL}^{cw} = \frac{\rho_{av}}{1500} D_{LL}^{cw} \qquad (3.9)$$

where ρ_{av} is the average density of the growth ring(s), and ρ_o the bulk density of the cell wall (1500 kg/m^3). Although based on the measured density of wood, ρ_{av} is an adjusted value that disregards the mass of mechanically ineffective components such as ray cells and extractives. Such adjustments are species dependent but will usually not amount to more than 10% of the unmodified density.

As in Section 3.2, individual cells can be modelled as having rectangular, circular or hexagonal cross-sections. Softwood cells are approximately hexagonal in cross-section and are arranged in essentially straight-line patterns in the radial direction (Figure 2.7). In the tangential direction cells have relatively irregular arrangement. The result is that when tangential force is applied radial cells walls tend to be loaded in bending, while when radial force is applied radial cell walls tend to be loaded axially (Figure 3.4). For hardwoods there is also more morphological order in the radial direction than in the tangential direction (Figure 2.8). Generalised statements about how cell walls behave are not reliable and should be ignored. Two-dimensional analytical solutions presume, for mathematical tractability, that in the $R-T$ plane cells within a growth ring(s) are uniform (Gillis, 1972; Koponen *et al.*, 1989, 1991; Gibson and Ashby, 1988). Cells are taken to be infinitely long tubes in the L direction, and they are assigned average geometric and stiffness properties. Such simplified models by their nature neglect contributions of cells other than those oriented in the L direction to estimates of strength or deformation characteristics of growth rings. This discrepancy versus reality is particularly important in the R direction because ray cells stiffen and strengthen the system. The notion of modelling based on average size cells is most

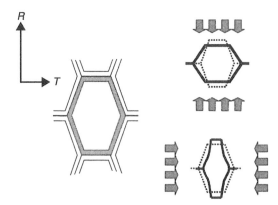

Figure 3.4   Hexagon representation of a softwood cell in the $R-T$ plane and deformation due to in-plane force.

obviously unreliable in the case of diffuse-porous or ring-porous hardwoods. Although average cell-based calculations may lead to reasonable estimates of bulk stiffness and shrinkage properties of some wood species, the approach is not well suited to estimation of strength properties of any species.

Cell dimensions are in reality variable in the $R - T$ plane within and between growth rings, with local pseudo-random variations between adjacent cells or adjacent clusters of cells. Failure processes on the meso-scale initiate at individual cells or clusters of cells due to local instability and presence of micro-defects. Thus, there is an inherent requirement in strength analysis to model behaviour at the growth ring level via numerical methods (Stefansson, 1995; Persson, 1997). Although finite element analysis techniques readily lend themselves to this, the resultant set of simultaneous equations can be so enormous that the method soon becomes impractical. A compromise approach is usually required. Such an approach is intermediate between continuum mechanics and a fully realistic modelling of wood features. A lattice network finite element model is a suitable tool for the meso, or indeed macro, scale. Such a model integrates morphology-based modelling with experimental techniques to identify micro-structural properties within two- or three-dimensional space. The material is discretised into a network of longitudinal beam elements, and transversal and diagonal spring elements that represent wood cells and their micro-structural connections. Micro-scale heterogeneity is implemented via Monte-Carlo simulation of statistical distributions of strength and stiffness for each type of component. The procedure by its nature introduces inherent disorder of the heterogeneous wood morphology. Lattice network models are discussed further in Section 7.4.3 in the context of fracture modelling.

3.4 Macro Scale: Clear Wood

3.4.1 *General considerations*

The macro-scale is the level at which wood is generally represented for the purposes of two- or three-dimensional stress analysis of engineered components. It is an acceptable level of representation for analysis of components in which stress variations in a direction are not large over distances about one order of magnitude greater than cell dimensions.

Clear wood is free from gross strength affecting features such as knots, grain deviation, resin pockets and irregularities in the growth ring structure. Whether gross features enhance or diminish apparent strength depends upon the property involved. For example, tension related axial strength properties are diminished due to presence of knots, but shear strength can be significantly enhanced by presence of knots. Properties are determined from specimens of limited dimension that represent bulk behaviour averaged over relatively few growth rings. Determination and specification of properties of clear wood is relative to orientation of primary cells such as tracheids in softwood and fibres in hardwood, with variations between the L, R and T directions being recognised. Property variation within the L, R and T directions is usually not addressed. In the presence of log taper and spiral grain properties labelled as belonging to an axis direction will not coincide with the global directions of the same names (L_o, R_o, T_o). Global directions are those defined relative to the axis in a tree stem

(Figure 3.2). Property determinations are based on the presumption that clear wood is homogenous and continuous.

Density has long been accepted as strongly related to physical and mechanical properties of clear wood of various species (Mullins and McKnight, 1981; USDA, 1989). An assumption of linear proportionality is often made, but relationships between mechanical properties and density are often not linear (Table 3.1). Apart from density, mechanical properties are strongly dependent on the moisture content of the wood at the time of loading. Traditional practice is to determine properties for wood in the green (saturated cell wall) condition and at 12% moisture content. In the case of static (monotonic short-term load) properties, interpolation procedures are used to estimate properties at moisture contents other than those at which those properties have been measured (Brown *et al.*, 1952):

$$\log X_3 = \log X_1 + [(m_1 - m_3)/(m_1 - m_2)] \log(X_2/X_1) \quad (m_2 \leq m_3 \leq m_1) \quad (3.10)$$

where X_3 is the predicted property, X_1 and X_2 are measured strength values, and m_i values are associated moisture contents. Usually X_1 is the green strength and m_1 the *FSP*. Values of *FSP* vary between species and can range from 24% to over 30%, but for the purposes of Equation (3.10) m_1 is taken to be 25%. In general there are important transition points in mechanical properties of wood at about $m = 6\%$ and 22%, and between these two points the relationship between mechanical properties and m is sensibly linear. Suppose, for example, it is required to estimate the compressive strength parallel to grain of Sitka Spruce (*Picea sitchensis* Carr.) at 18% moisture content, and measured properties are $X_1 = 17.7$ MPa at $m_1 = 25\%$, and $X_2 = 37.8$ MPa at $m_2 = 12\%$. For $m_3 = 18\%$ Equation (3.10) yields: $\log X_3 = \log 17.7 + [(25 - 18)/(25 - 12)] \log(37.8/17.7) = 1.425$, and thus $X_3 = 26.6$ MPa. This compares with 27.0 MPa as determined by simple linear interpolation. Some caution should

Table 3.1   Functions relating some mechanical properties of clear wood to relative density *RD* (based on Mullins and McKnight, 1981)

Property (MPa)	Relative density-property relationship	
	Green wood	Air-dry (12% moisture content)
MOE in bending, $E_{L,b}$	$E_{L,b} = 16\,300\ RD$	$E_{L,b} = 19\,300\ RD$
MOR in bending, $f_{L,b}$	$f_{L,b} = 121\ RD^{1.25}$	$F_{L,b} = 177\ RD^{1.25}$
MOE in compression parallel to grain, $E_{L,c}$	$E_{L,c} = 20\,100\ RD$	$E_{L,c} = 23\,300\ RD$
Compressive strength parallel to grain, $f_{L,c}$	$f_{L,c} = 46.4\ RD$	$f_{L,c} = 84.1\ RD$
Compressive strength perpendicular to grain, $f_{P,c}$	$f_{P,c} = 20.7\ RD^{2.25}$	$f_{P,c} = 31.9\ RD^{2.25}$

Values predicted are average properties as determined in short duration static tests (rate of loading producing failure in less than about 0.1 hours). *RD* is calculated using oven-dry mass and volume at the moisture condition indicated. *MOE* is modulus of elasticity. *MOR* is modulus of rupture = apparent strength assuming a homogenous elastic response. Differences between $E_{L,b}$ and $E_{L,c}$ reflects in part that $E_{L,b}$ is not a 'shear free' estimate of modulus of elasticity.

be exercised in using published strength properties of clear wood in dry condition and Equation (3.10), as they apply only to 'normal' wood dried slowly at moderate temperature to the moisture content at the time of testing.

Properties such as transverse shrinkage, static strength and stiffness in the longitudinal direction depend strongly on the microfibril angle, *MFA* (Cave, 1969; 1976; Alkan, 1986; Cave and Walker, 1994; Megraw *et al.*, 1999). For *Eucalyptus delegatensis* the combination of density and *MFA* has been found to account for up to 96% of variation in dynamic modulus of elasticity in the longitudinal direction, $E_{L,dyn}$ with *MFA* alone accounting for up to 86% of the variation (Evans and Ilic, 2001). For Red pine density and *MFA* accounts for 85% of variation in $E_{L,dyn}$, with *MFA* alone accounting for 79% of variation in $E_{L,dyn}$ (Alkan, 1986). It has been speculated that E_L does not depend upon density at all, and instead that density reflects a correlation with factors such as proportions, spatial distributions and states of cellulose and lignin (Evans and Ilic, 2001). Likewise, strength properties are probably not physically dependent upon density.

The relationship has been established (Evans and Ilic, 2001):

$$E_{L,dyn} = 0.284 \frac{D_{ss}}{s_{002}} \quad \text{GPa} \qquad (3.11)$$

where D_{SS} is gravimetric density and s_{002} the standard deviation of the 002 azimuthal diffraction profile as determined by x-ray diffractometry. The relationship exists $MFA \approx 1.28\sqrt{s_{002}^2 - 36}$. Equation (3.11) has also been applied to *Pinus radiata* and other species and it is speculated that it might well be universally valid (Evans *et al.*, 2000). The drawback is that measurement of s_{002} requires equipment not commonly found in wood testing laboratories.

Structural analysts commonly presume property values at 20°C are applicable, but appropriateness of this needs careful consideration, except perhaps if a structural system is covered and in a location with a temperate climate. An increase in temperature relative to normal room conditions tends to reduce mechanical properties of wood, and a reduction in temperature tends to increase them. Such effects are not independent of the moisture content. This can be demonstrated by considering wood in the green condition. If free water in cell lumen is frozen it will have the effect of stiffening and strengthening the wood, but obviously any gain is lost immediately upon thawing and the wood may actually have been damaged in the process. For positive temperatures below 100°C effects of rapid changes in temperature are essentially reversible. Moderately high temperature combined with high level compressive stress (hot-pressing) modifies wood structure and residual properties, a behaviour that is exploited in manufacture of many types of wood-with-glue composites. A study by Tabarsa and Chui (1996) on hot-pressing of white spruce showed that residual mechanical properties change rapidly for press temperatures between 100°C and 150°C, but not for press temperatures between 150°C and 200°C. Data on the combined effects of temperature and moisture has been summarised by the USDA (1989).

Irreversible degradation of the wood substance occurs at elevated temperature or at moderate temperature if sustained. Degradation of wood results in weight loss and reduction in mechanical properties. Strength and stiffness loss is proportional to temperature, exposure period, the heating medium, moisture content, member size and wood species. Proportional loss of residual strength tends to be greater than proportional loss of residual stiffness, with reductions being proportional to the cumulative exposure time (USDA, 1989). Long-term heating effects on the strength of wood are negligible at temperatures below about 50°C, and thus irreversible degradation of wood is not normally an issue in structural applications. Exceptions occur when wood elements are located in industrial structures such as cooling towers, in roof spaces of residential construction in hot regions, or walls where abutting chimneys are not properly insulated.

Tree stems are regarded as cylindrically orthotropic with properties defined in the $L - R - T$ Cartesian co-ordinate system. However, when there is taper in the log and spiral grain, longitudinal wood cells do not lie in the direction of the axis of the log (L_o). It is necessary therefore to be able to transform properties between local and global co-ordinate systems (Figure 3.5). Transformations of properties between local and global co-ordinates systems are made by standard mechanics techniques based on direction cosines.

In a broad sense there is correlation between strength and stiffness properties of clear wood, because both are related to the amount of wood substance in cell walls and the porosity which reflects in density. Figure 3.6 shows the relationship between modulus

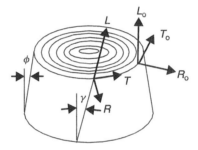

Figure 3.5   Relationship of local $L - R - T$ Cartesian co-ordinate system relative to global $L_o - R_o - T_o$ Cartesian co-ordinate system (γ = spiral growth angle, ϕ = angle of taper).

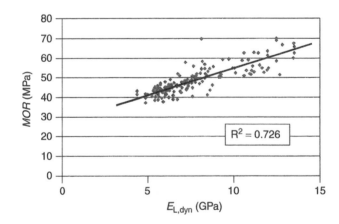

Figure 3.6   Relationship between *MOR* and $E_{L,dyn}$ for a single Red pine at 12% moisture content (drawn using data by Alkan, 1986).

of rupture, *MOR* (apparent bending strength), and dynamic longitudinal modulus elasticity, $E_{L,dyn}$, for Red pine. As can be seen, there is significant residual deviation about the general trend. The case illustrated is typical of the quality of such correlation. Fitted relationships are therefore only useful as rough guides to likely strength properties of particular pieces of clear wood. Theoretically, correlation of strength to stiffness relates to the homogeneity of a material (Nielsen, 1997). As homogeneity of wood varies in the *R* direction, and to a lesser extent in the *L* direction, it follows that the quality of any correlation will vary between individual trees and between species. This is why non-destructive evaluation techniques that ignore the physical structure of individual subject pieces are ineffective for wood, and other variable porous materials.

3.4.2 Elastic response

Wood exhibits nonlinear inelastic strain at high stress levels, or if stress is sustained. Here stress level *SL* is taken to be the applied stress as a proportion of the corresponding strength as measured in a static test. However, for loading applied for less than about 0.1 hours and a low *SL* the response of clear wood can be assumed to be linear elastic. Locally curvature in growth rings is ignored. In a Cartesian sense, clear wood is thought of as orthotropic material. Assuming orthotropy, Figure 3.7 illustrates stress-strain behaviour under various short-duration stress states as a function of stressing directions *L*, *R* and *T*. Apparent elastic properties as determined in static tests are strongly directionally dependent (Table 3.2). Elastic moduli for *R* and *T* directions are much lower than the elastic modulus in the L direction. Property ratios and Poisson's ratios can vary significantly between species. Bodig and Goodman (1973) summarise apparent elastic constants for a broad range of softwood and hardwood species of commercial importance. Those authors provide best-fit power law regression models

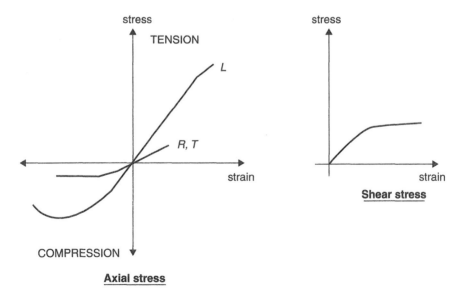

Figure 3.7 Stress-strain responses of clear wood as a function of stressing direction.

Table 3.2   Ratios of apparent elastic moduli for clear wood in dry condition (based on Bodig and Goodman, 1973; UDSA, 1989)

Species	Relative density RD	Moisture content m (%)	E_T/E_L	E_R/E_L	G_{LR}/E_L	G_{LT}/E_L	G_{RT}/E_L
Balsa	0.13	9	0.015	0.046	0.054	0.037	0.005
Spruce	0.37	12	0.041	0.074	0.050	0.061	0.002
Yellow-poplar	0.38	11	0.043	0.092	0.075	0.069	0.011
Douglas-fir	0.50	12	0.050	0.068	0.064	0.078	0.007
Mahogany	0.50	12	0.073	0.107	0.098	0.066	0.028
Sweetgum	0.53	11	0.050	0.115	0.089	0.061	0.021
Black Walnut	0.59	11	0.056	0.106	0.085	0.062	0.021
Alpine Maple	0.59	10	0.088	0.152	0.123	0.110	0.029
Yellow Birch	0.64	13	0.050	0.078	0.074	0.068	0.017

Table 3.3   Poisson's ratios for clear wood in dry condition (based on Bodig and Goodman, 1973; UDSA, 1989)

Species	Relative density	Moisture content m (%)	ν_{LR}	ν_{LT}	ν_{RT}	ν_{TR}	ν_{RL}	ν_{TL}
Balsa	0.13	9	0.23	0.49	0.67	0.23	0.02	0.01
Spruce	0.37	12	0.44	0.56	0.57	0.29	0.03	0.01
Yellow-poplar	0.38	11	0.32	0.39	0.70	0.33	0.03	0.02
Douglas-fir	0.50	12	0.29	0.45	0.39	0.37	0.04	0.03
Mahogany	0.50	12	0.31	0.53	0.60	0.33	0.03	0.03
Sweetgum	0.53	11	0.32	0.40	0.68	0.31	0.04	0.02
Black Walnut	0.59	11	0.50	0.63	0.72	0.38	0.05	0.04
Alpine Maple	0.59	10	0.46	0.50	0.82	0.40	0.09	0.04
Yellow Birch	0.64	13	0.43	0.45	0.70	0.43	0.04	0.02

ν_{ij} is strain in the j direction due to unit strain in the i direction.

to relationships between elastic or shear moduli and density, or between other moduli and E_L. Separate models are given for the softwood and hardwood species that were considered. Available evidence shows no correlation between Poisson's ratios and density (Table 3.3).

Although for practical purposes wood can often be regarded as brittle under static tension, it is highly non-linear and apparently ductile in behaviour under static compression or shear. Apparent ductility means that the bulk stress-strain response under monotonic short-term loading has a shape similar to that observed for metals, with no intent to imply that there is actually plastic deformation. Wet or green wood can exhibit non-linear behaviour at much lower stress levels than observed for dry wood.

Within the qualifications just mentioned and adopting the $L - R - T$ Cartesian co-ordinate system, the generalised Hooke's law for orthotropic material is:

$$\varepsilon^e = C \, \sigma \tag{3.12a}$$

or

$$
\begin{Bmatrix} \varepsilon_{LL} \\ \varepsilon_{RR} \\ \varepsilon_{TT} \\ \gamma_{LR} \\ \gamma_{LT} \\ \gamma_{RT} \end{Bmatrix} = \begin{bmatrix} \dfrac{1}{E_L} & \dfrac{-\nu_{RL}}{E_R} & \dfrac{-\nu_{TL}}{E_T} & 0 & 0 & 0 \\ \dfrac{-\nu_{LR}}{E_L} & \dfrac{1}{E_R} & \dfrac{-\nu_{TR}}{E_T} & 0 & 0 & 0 \\ \dfrac{-\nu_{LT}}{E_L} & \dfrac{-\nu_{RT}}{E_R} & \dfrac{1}{E_T} & 0 & 0 & 0 \\ 0 & 0 & 0 & \dfrac{1}{G_{LR}} & 0 & 0 \\ 0 & 0 & 0 & 0 & \dfrac{1}{G_{LT}} & 0 \\ 0 & 0 & 0 & 0 & 0 & \dfrac{1}{G_{RT}} \end{bmatrix} \begin{Bmatrix} \sigma_{LL} \\ \sigma_{RR} \\ \sigma_{TT} \\ \tau_{LR} \\ \tau_{LT} \\ \tau_{RT} \end{Bmatrix} \tag{3.12b}
$$

where ε^e is the elastic strain vector, C is the elastic compliance matrix, σ is the stress vector; E_L, E_T, E_R are elastic moduli in L, T and R directions; G_{LT}, G_{TR} and G_{RL} are shear moduli for planes $L-T, T-R$ and $R-L$; and ν_{ij} are Poisson's ratio $(i, j = L, T, R)$. To a good approximation, it can be assumes that $\nu_{ij}/E_i = \nu_{ji}/E_j$, which reduces the number of independent constants from twelve to nine. This makes the compliance matrix symmetric, thereby greatly simplifying analysis.

The inverse relationship to Equation (3.12) is:

$$
\sigma = D\varepsilon^e = C^{-1}\varepsilon^e \tag{3.13}
$$

where D is the elastic stiffness matrix.

In the linear elastic range strains associated with externally applied stress, moisture induced shrinkage or swelling, and temperature are additive. Thus:

$$
\varepsilon = \varepsilon^e + \varepsilon^m + \varepsilon^{Tmp} \tag{3.14}
$$

where
$$
\varepsilon^m = \begin{bmatrix} \alpha^L \\ \alpha^T \\ \alpha^R \\ 0 \\ 0 \\ 0 \end{bmatrix} \Delta m, \quad \varepsilon^{Tmp} = \begin{bmatrix} \lambda^L \\ \lambda^T \\ \lambda^R \\ 0 \\ 0 \\ 0 \end{bmatrix} \Delta Tmp
$$

$\varepsilon, \varepsilon^e, \varepsilon^m, \varepsilon^{Tmp}$ are total, external stress, moisture change, and temperature change induced strains; α^i values are shrinkage coefficients; and λ^i values are thermal expansion coefficients. Neutral points for moisture content m and temperature Tmp must be defined by the analyst. For example, the neutral point for m might be defined as the *FSP* or an in-service equilibrium value, and the neutral point for Tmp as 20°C. Taking the neutral point for m as *FSP* yield strains that include the effect of drying to a moisture content below the green state.

3.4.3 In-elastic response

Macro responses illustrated in Figure 3.7 for clear wood reflect that it is not in fact a continuous material. The cellular and porous nature of wood permits substantial compaction under compression and shear. Transverse expansion is strongly influenced by the amount of strain in the longitudinal direction and Poisson's ratios only apply when elasticity prevails (Gong and Smith, 2000). Piecewise incremental methods have been used to represent the non-linear response of fibre composites (Chiang, 1983), and nonlinear compressive and shear stiffness of wood has been represented by trilinear stress-strain relationships (Patton-Mallory *et al.*, 1997). However, piecewise incremental methods ignore coupling between material behaviour in directions of material symmetry and do not obey constitutive laws for continuous materials (Kharouf, 2001). Such methods result in lack of numerical convergence and predict regions of unrealistically high stress. The exception to this is if only the shear deformation is non-linear. Incremental solution methods are successful in that instance because shear is an uncoupled behaviour (Petit and Waddoup, 1969; Hashin *et al.*, 1974; Sandhu, 1976).

Modelling wood as elasto-plastic material avoids violation of constitutive laws for continuous materials. The approach has proved successful when wood is loaded in confined or semi-confined compression (Francois, 1992; Moses, 2000; Kharouf, 2001). The best known application of plasticity theory to wood is in limit analysis of joints with dowel-type fasteners (Johansen, 1949), in which the plastic yield capacity is predicted. Permanent deformation in compression can be described macroscopically within the framework of plasticity theory despite the previously mentioned lack of physical correctness.

In plasticity theory, a set of constitutive equations for a multi-axial stress state is derived from uni-axial stress-strain data. This is accomplished through a yield criterion, flow rule and hardening rule (Chen and Han, 1988). Hill (1947, 1950) postulated a yield surface for orthotropic material as an extension of the Von-Mises criterion for isotropic materials. Isotropic hardening is presumed which leads to proportional change in orthotropic parameters during hardening. The Hill criterion has been generalised to include non-proportional hardening (Whang, 1969), differences in strength under tension and compression (Shih and Lee, 1978), updated yield surfaces (Gotoh, 1977) and softening behaviour (Louranco *et al.*, 1997). Elasto-plastic modelling of wood has been performed for bi-axial compression (Kharouf, 2001) and tri-axial stress states (Moses, 2000) based on the Hill yield criterion.

For plastically anisotropic materials the yield condition is given by (Shih and Lee, 1978):

$$f = (\sigma_{ij}, \alpha_{ij}, A_{ijkl}, k) = 0 \tag{3.15}$$

where σ_{ij} is the second order stress tensor, $A_{ijkl}(i, j, k, l = 1, 2, 3)$ denotes the fourth order tensor of anisotropic strength parameters describing the shape of the yield surface, α_{ij} describes the origin of the yield surface, and k is a scalar parameter representing the reference yield stress.

In plane stress, and assuming a transversely isotropy, the Hill criterion for bi-axial compression is:

$$f = \left(\frac{\sigma_1}{\sigma_{1c}}\right)^2 + \left(\frac{\sigma_2}{\sigma_{2c}}\right)^2 + \left(\frac{\tau}{S}\right)^2 - \frac{\sigma_1 \sigma_2}{\sigma_{1c}^2} - 1 = 0 \tag{3.16}$$

where σ_i is compressive stress in the axis direction i, σ_{ic} is compressive strength in the axis direction i, τ is shear stress, and S is shear strength. It is usually assumed that axis 1 is parallel to grain and axis 2 perpendicular to grain. For consistency, during plastic flow the state of stress on the yield surface must result in:

$$df = 0 \qquad (3.17)$$

The total strain increment in the plastic regime is:

$$\{d\varepsilon\} = \{d\varepsilon^P\} + \{d\varepsilon^e\} = \{d\varepsilon^P\} + [C]\{d\sigma\} \qquad (3.18)$$

where $\{d\varepsilon^P\}$ and $\{d\varepsilon^e\}$ are plastic and elastic components respectively, and $[C]$ the elastic compliance matrix, Equation (3.12). The flow rule determines the direction of plastic straining and is given as:

$$d\varepsilon^P = d\lambda \frac{\partial f}{\partial \sigma} = d\lambda a \qquad (3.19)$$

where $d\lambda$ is a plastic multiplier, which determines the amount of plastic straining, and $\partial f / \partial \sigma = a$ is termed the flow vector. The anisotropic hardening rule by Vaziri *et al.* (1993) was applied by Kharouf (2001), which allows for non-proportional change in yield values and thus non-uniform expansion of the yield surface during plastic flow. The constitutive equations are:

$$d\sigma_{ij} = D_{ij}^{ep} d\varepsilon_j \qquad (3.20)$$

where
$$D_{ij}^{ep} = D_{ij} - \frac{D_{ik} a_k a_l D_{lj}}{\left(1 - \frac{1}{2k}\sigma_i \sigma_j \frac{\partial A_{ij}}{\partial k}\right) H' + a_m D_{mn} a_n}$$

D is the elastic stiffness matrix, D^{ep} the elasto-plastic stiffness matrix, and H' the plastic modulus based on effective stress versus effective strain. Perfect plasticity, hardening or softening, is modelled depending on the value of H'.

Moses (2000) applied the Hill criterion to wood and wood-strand based composites subject to tri-axial stress states. He incorporated the work hardening model by Valliappan *et al.* (1976) and assumed an associative flow rule. Apart from elastic compliance terms, the model requires bi-linear stress-strain curves representing tension and compression response in each direction of material symmetry and shear in the three planes of material symmetry (Figure 3.8). This means 18 experimentally determined independent plastic constants are required. Plastic incompressibility requires:

$$\frac{\sigma_{1t} - \sigma_{1c}}{\sigma_{1t}\sigma_{1c}} + \frac{\sigma_{2t} - \sigma_{2c}}{\sigma_{2t}\sigma_{2c}} + \frac{\sigma_{3t} - \sigma_{3c}}{\sigma_{3t}\sigma_{3c}} = 0 \qquad (3.21)$$

where σ_{it} and σ_{ic} are yield stresses in tension and compression respectively for direction i. The 1-axis direction is taken to be the L direction. Maintaining a closed yield surface, to avoid numerical instability, requires that:

$$M_{11}^2 + M_{22}^2 + M_{33}^2 - 2(M_{11}M_{22} + M_{22}M_{33} + M_{11}M_{33}) < 0 \qquad (3.22)$$

where $M_{ii} = \frac{K}{\sigma_{it}\sigma_{ic}}$ $(i = 1, 2, 3)$, $K = \sigma_{1t}\sigma_{1c}$.

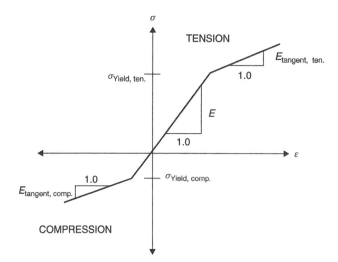

Figure 3.8 Non-symmetric bilinear stress-strain curve for anisotropic plasticity model.

Satisfying Equation (3.22) at all levels of strain is reportedly difficult, and it is necessary to employ numerical optimisation strategies to define suitable estimates of tangent stiffness values that define hardening (Moses, 2000).

As previously noted, the notion that unmodified wood has an elasto-plastic response when loaded in tension is not valid, however the same cannot be automatically presumed for wood-based composites made with synthetic resins.

3.4.4 Strength criteria

Various strength criteria have been proposed for predicting localised material failure due to multi-axial stress caused by static load. Criteria are usually applied in terms of a stress space, but can also be in terms of a strain space. Typically, they are expressed in polynomial form. The generalised form of the failure surface for orthotropic material is (Tsai and Wu, 1971):

$$f(\sigma_k) = F_i \sigma_i + F_{ij} \sigma_i \sigma_j = 1.0 \quad (i, j, k = 1, 2, \dots, 6) \tag{3.23}$$

where F_i and F_{ij} are strength tensors of second and fourth rank, respectively. Simplified forms have been proposed because the general form is too complex for practical use. Differences in tensile and compressive strengths in the directions of material symmetry are accounted for in the strength tensors (e.g. Equations (3.25) and (3.26). Bi-axial criteria have been used for wood (Norris, 1962; Cowin, 1979; Liu, 1984; Rowlands, 1985; Rahman *et al.*, 1991), with the most common being:

Norris (1962): $$\sqrt{\left(\frac{\sigma_x}{X}\right)^2 + \left(\frac{\sigma_y}{Y}\right)^2 + \left(\frac{\tau_{xy}}{S}\right)^2} = 1.0 \tag{3.24}$$

Tsai-Wu (1971): $$F_1 \sigma_x + F_2 \sigma_y + 2F_{12} \sigma_x \sigma_y + F_{11} \sigma_x^2 + F_{22} \sigma_y^2 + F_{66} \tau_{xy}^2 = 1.0 \tag{3.25}$$

$$F_1 = \frac{1}{X} - \frac{1}{X'}, F_2 = \frac{1}{Y} - \frac{1}{Y'}$$

$$F_{11} = \frac{1}{XX'}, F_{22} = \frac{1}{YY'}$$

$$F_{66} = \frac{1}{S^2}, F_{12} \leq \sqrt{F_{11}F_{22}}$$

Cowin (1979):   Same as Equation (3.25), except that $F_{12} = \sqrt{F_{11}F_{22}} - \frac{1}{2S^2}$
(3.26)

X and Y are tensile strengths in the x and y directions; X' and Y' are compressive strengths in the x and y directions; and S is shear strength in the xy plane. If at any point in a component the summation of terms, in an equation such as (3.24)–(3.26), is less than 1.0 the material is predicted to be undamaged. The Tsai–Wu and Cowin criteria provide continuous functions between compression-compression or tension-tension regions and compression-tension regions of the stress space. Simpler criteria such as that due to Norris have to be applied in a piecewise manner if it is required to model more than the tension-tension region. Strength properties entering any criterion are determined through tests on small specimens of clear wood. Some caution is required when determining X, Y, X', Y', and S as standardised test methods used to quantify such properties usually do not produce pure stress states and have different reference volumes.

Although theoretical arguments are advanced in support of the various strength criteria, they are all empirical in nature. It is important to recognise that no strength criterion can be expected to work successfully in regions with steep stress gradients, because they all assume that the material is homogenous and continuous. Strength criteria are incapable of representing stress redistribution that occurs in wood at meso and lower scales, or variability in properties within and across growth rings. Appropriate use of strength criteria is prediction of the location(s) at which a fracture plane(s) in wood will develop (Tan and Smith, 1999), rather than as tools for predicting the level of externally applied load at which fracture will occur. Strength criteria such as those mentioned should only be expected to yield hand-waving estimates of failure loads.

Figure 3.9 illustrates a typical piecewise application of the Norris criterion, Equation (3.24), to produce a bi-axial failure surface. It can be seen that the effect of $|\tau_{xy}| \geq 0.0$ is to shrink the failure surface. Other criteria produce broadly similarly shaped failure surfaces. Differences between failure criteria tend to be small relative to uncertainty caused by scatter in experimental data. It is not possible to clearly state that one criterion is superior to others.

3.4.5   *Rheological behaviour*

Wood is a rheological material, and thus exhibits deformation that is dependent upon the loading history and total elapsed time. More generally, rheological behaviour is also a function of the thermal and moisture histories, and their interaction with the loading history. Creep (flow) rates increase under sustained or fluctuating elevated temperatures (Dinwoodie *et al.*, 1991). For construction applications of wood, predictions of creep

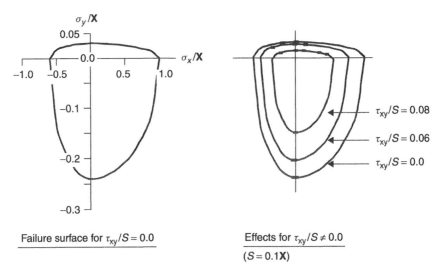

Figure 3.9 Typical failure surface based on the Norris strength criterion: $X' = 0.6X$, $Y = 0.03X$, $Y' = 0.24X$.

usually concern behaviour at temperature well within the range where wood behaves as a stable solid. At temperatures less than 50°C the influence of temperature is much less than that of moisture, and can usually be ignored. Consideration is normally given to the effect of stress alone (creep), or stress combined with moisture (mechanosorptive creep). However, as air temperature can be quite high, in for instance domestic roof spaces or under industrial conditions, it is not always appropriate to ignore temperature when predicting deformation. Scientifically, creep refers to deformation behaviour under constant climatic conditions. When test conditions, and thus moisture, are only loosely controlled it is proper to talk of time-dependent, rather than creep, behaviour. Such a distinction is adhered to below.

Like other engineering materials, wood exhibits three creep phases under constant stress (Figure 3.10). Primary creep is an initial phase during which the rate of deformation accumulation decreases with any increase in elapsed time. Secondary creep is a phase during which the rate of deformation accumulation is constant. Tertiary creep is a phase during which the rate of deformation accumulation increases with any increase in elapsed time. Whether secondary or tertiary creep phases are entered depends upon the stress level and the elapsed time. The terms 'creep rupture' and 'static fatigue' are synonymous and refer to a situation where tertiary deformation has progressed to the point of specimen failure. Literature on wood often uses the expression 'duration of load effect on strength' in respect of situations where time-dependent deformation propagates to a critical tertiary condition (catastrophic loss of strength). Static fatigue/duration of load effect on strength is dealt with at length in Chapters 6 and 8, and therefore is not considered further in this chapter.

Deformation (strain) at any time following loading is thought of as consisting of elastic and delayed components. It is usually presumed that if load is instantaneously removed wood will experience instantaneous recovery of elastic strain equal in magnitude to elastic strain experienced when that load was applied. This is a reasonable

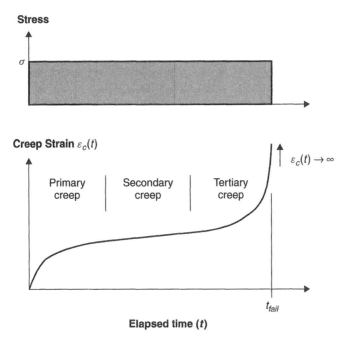

Figure 3.10 Possible deformation phases in creep tests.

presumption when the *SL* is low. The delayed strain has irrecoverable viscous, and delayed recoverable components (Figure 3.11). There is always some residual strain irrespective of how long the recovery period following removal of stress. The creep response of wood is strictly non-linear with respect to *SL* (Bach, 1965; Whale, 1988) (Figure 3.12). The term 'relative creep' *RC* is the delayed strain component expressed as a proportion of instantaneous strain ($RC = \{\varepsilon(t) - \varepsilon_o\}/\varepsilon_o$). To a reasonable approximation, creep response of wood can be regarded as linear if the *SL* is less than about 0.35 (Andriamaitantosa, 1995). Relatively simple models can be used when the stress level is low, but modelling of creep in wood subjected to high stress levels must explicitly recognise *SL*-dependent nonlinearity.

As shown in Figure 3.13, relative creep *RC* depends strongly upon the moisture conditions at the commencement of and during loading. *RC* is strongly related to the moisture content of wood at the time of loading, and any change in the moisture content while under load (Armstrong and Kingston, 1962). The greatest influence on *RC* is drying out of wood while under load. This has been explained as the result of lowered potential energy barriers against molecular slip during drying processes (van der Put, 1987a). Although mechanosorptive effects on creep in clear wood have been observed frequently, they are not understood well.

All creep models are phenomenological and need to be calibrated from experimental data. Under conditions where creep behaviour is sensibly linear with respect to *SL*, and where behaviour never progresses beyond the secondary phase, the well known Burger Body 4-element dash-pot and spring model (or similar models with a few extra elements) yield reasonable indications of expected levels of strain (Figure 3.11). For

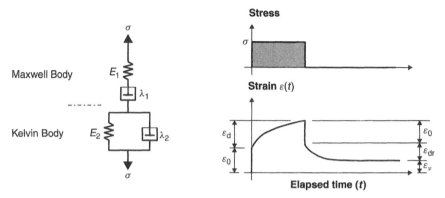

Figure 3.11   Burger Body model.

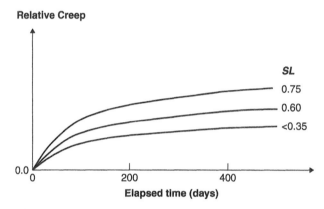

Figure 3.12   Effect of stress level on relative creep in wood (schematic).

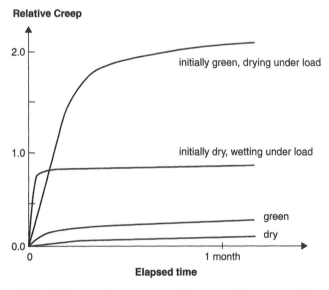

Figure 3.13   Influence of moisture condition on relative creep (schematic).

constant stress σ the Burger Body predicts strain at an elapsed time of t to be:

$$\varepsilon(t) = \sigma \left(\frac{1}{E_1} + \frac{t}{\lambda_1} + \frac{(1 - e^{-(E_2/\lambda_2)t})}{E_2} \right) \qquad (3.27)$$

where E_1 and E_2 are spring constants; and λ_1 and λ_2 are coefficients of traction, each calibrated from experimental data. The term σ/E_1 is the instantaneous recoverable strain ε_o (Figure 3.11). Just prior to when stress is removed total strain consists of ε_o and delayed irrecoverable viscous creep ε_v (Maxwell Body), and delayed recoverable strain ε_{dr} (Kelvin Body). In practice, definition of ε_o is arbitrary, because stress can never actually be applied instantaneously and because there is no delay prior to creep. Some definition of ε_o must be picked (e.g. strain 1 minute, 1 hour or 1 day after attainment of the target *SL*). The preferred definition of ε_o often is that which leads to a good fit of the model during primary and secondary creep phases. Analysts are quite variable in their choice of ε_o.

More complex discrete element mechanical analogue (spring and dash-pot) models have been developed treating wood as behaving like a system of parallel Maxwell elements (Figure 3.14). The most complex employ non-linear dash-pots that respond to externally applied stress, temperature and moisture, based on the theory of molecular deformation kinetics (van der Put, 1987b).

Generalised (as opposed to discrete element) models fall in two broad categories:

- Multiple integral series representations;

- Single integral representations.

Multiple integral theories embody the concept of 'material memory' via kernel functions that encompass more than one time argument.[1] This accounts for nonlinear interactions between past and present loading events. Although theoretically compelling, multiple integral theories are usually too difficult to calibrate. Simpler single

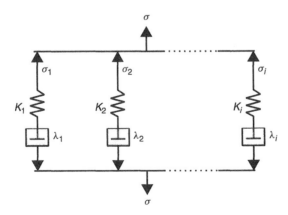

Figure 3.14 Multiple Parallel Maxwell Body model.

---

[1] Kernel functions are time-dependent functions that relate observed strain with applied stress.

integral alternatives involve only a linear kernel function, with various nonlinear super-position strategies (e.g. combinations of linear kernels, implicit incorporation of a stress level dependency within the kernel function, use of a 'reduced' instead of a real time argument within the kernel function). A hybrid model developed by Whale (1988) is arguable the most general developed with behaviour of wood in mind. Although it was developed to represent the apparent compressive strain response of wood loaded parallel to grain by a cylindrical metal dowel, the model could be applied to any uni-axial stress state. It can handle irregular sustained stress, but not stress reversal. The model takes the form:

$$\varepsilon(t) = J_0 \sigma(t) + \int_{\tau=0}^{\tau=t} J_f(\sigma^*(\tau), \psi(t)-\psi(\tau)) g_1 \left[\frac{\sigma^*(\tau)}{\partial \sigma^*(\tau)/\partial \tau} \right] \frac{\partial \sigma^*(\tau)}{\partial \tau} d\tau$$

$$+ \int_{\tau=0}^{\tau=t} J_c(t-\tau) \frac{\partial g_2[\sigma(\tau)]}{\partial \tau} d\tau \qquad (3.28)$$

where

$$\psi(t) = \int_{s=0}^{s=t} a_\sigma \left(\frac{\sigma(s)}{\sigma^*(\tau)} \right) ds \, ; \, \psi(\tau) = \int_{s=0}^{s=\tau} a_\sigma \left(\frac{\sigma(s)}{\sigma^*(\tau)} \right) ds$$

J_0 is the elastic compliance; J_f is the flow compliance; J_c is small stress creep compliance; $\sigma^*(\tau)$ is the highest stress level existing at τ; g_1 is a nonlinear function of the ratio between the new highest *SL* and the change in the highest *SL*; g_2 is a nonlinear function of stress; a_σ is a nonlinear function of stress; and τ and s are generic times. Inclusion of the memory component g_1 is crucial to modelling creep when there is an irregular stress history (Whale, 1988; Gong and Smith, 2003). Such a term needs to accounts for sequencing of stress levels in a load history. It is worth noting that Whale considered abilities of eleven creep models, apart from Equation (3.28), to predict displacement (strain) versus time responses for nail embedment specimens with pseudo irregular stress histories having peak *SL* values up to 0.45, and of up to 35 days duration. In most cases, model predictions did not match observations well after several changes in *SL*. There was no inherent correlation between model complexity and quality of predictions, which emphasises the need for careful model selection and verification of predictive capabilities.

There is no reason why complex phenomenological models should not be developed for cases where there is stress reversal, bi-axial or tri-axial stress states, or generalised rheological conditions (mechanosorptive or thermo-mechanosorptive response). However, the amount of data for calibration and validation would be very large, and attainment of robust calibration procedures very difficult.

3.5   Massive Scale: Structural Wood

3.5.1   *General considerations*

Lumber is the most common type of structural wood 'material'. Here it is taken to mean prismatic elements of sawn wood having a smallest cross-section dimension less than 100 mm. Lumber contains strength reducing features such as knots, limited deviation in the general-slope-of-grain and limited distortion of the geometry after sawing and drying processes. The extent of imperfections is linked to end-use related grade designations and engineering design properties. Massive wood means large dimension timbers produced by sawing (smallest cross-section dimension more than 100 mm), and solid wood members created by gluing together relatively small pieces of lumber (laminates) to create glued-laminated-timbers (glulam members). For the purposes of this discussion glulam is regarded as massive wood because the volumetric proportion of glue is very small. Massive wood products are given end-use related grade designations based on physical and/or mechanically evaluated features. Engineering design properties reflect grade distinctions. Apart from lumber and massive wood elements, various proprietary reconstituted 'wood with glue' substitute products are used for construction purposes, e.g. Laminated Veneer Lumber (LVL), Parallel Strand Lumber (PSL), and plywood. Because of the inherent presence of imperfections, structural wood has lower strength than clear wood of the same species. Imperfections in structural wood such as gaps, joints and overlaps, and lack of adhesive, are loci for stress concentrations and potential failure sites. Thus, under externally applied loading, the bulk of material in an element is relatively lightly stressed and will not contribute to failure initiation. Failure in structural wood relates to localised processes, and this is why knowledge of how wood behaves on the macro and lower scales is essential to scientific understanding of failure by fracture and fatigue processes.

3.5.2   *Representation of mechanical behaviour*

Almost universally, the approach to quantifying the strength and stiffness of structural wood is to test representative material from production with use of full-size and in-grade test specimens. 'Properties' of structural wood quoted in the literature are apparent rather than true values. Apparent properties presume linear elastic response and ignore heterogeneity. Although this is scientifically unrealistic, the approach permits engineers to make simple calculations that are acceptable from the standpoints of safety, serviceability and economy for everyday situations, with acceptability having been proven through practical experience.

For lumber and massive wood, elastic analysis gives recognition to global parallel (L_o) and perpendicular ($R_o - T_o$ plane) to grain directions, as defined in Figure 3.5. Taking the combined (perpendicular to grain) direction as axis 2, and the parallel to grain direction as axis 1 the constitutive equation (Equation (3.12)) reduces to:

$$\begin{Bmatrix} \varepsilon_{11} \\ \varepsilon_{22} \\ \gamma_{12} \end{Bmatrix} = \begin{bmatrix} \dfrac{1}{E_1} & \dfrac{-\nu_{21}}{E_2} & 0 \\ \dfrac{-\nu_{12}}{E_1} & \dfrac{1}{E_2} & 0 \\ 0 & 0 & \dfrac{1}{G_{12}} \end{bmatrix} \begin{Bmatrix} \sigma_{11} \\ \sigma_{22} \\ \tau_{12} \end{Bmatrix} \tag{3.29}$$

where E_1, E_2 are elastic moduli in directions 1 and 2; G_{12} is shear modulus for plane 1–2; ν_{ij} is Poisson's ratio. How realistic the assumption of transverse isotropy is depends upon species (Table 3.2). It should be borne in mind that pith to bark and longitudinal variations in elastic properties are often greater than localised differences in E_T and E_R. Lay-up pattern has a homogenising effect in the case of engineered wood products such as LVL, PSL and glulam. However, the effect of growth ring curvature on stress distributions can be significant for glulam members (Larsen, 2003), which will reflect itself in low apparent strength properties.

Time-dependent deformation behaviour of softwood lumber and glued-laminated-timber have been shown to be essentially equivalent when the moisture content does not exceed 20% (Rouger *et al.*, 1990). Under heated indoor climate conditions, permanently sustained load does not usually produce time-dependent deformation in structural size wood members that exceeds the elastic deformation under that load. However, if climate conditions result in a moisture content greater than about 20% significantly higher levels of time-dependent deformation will result. The influence that variations in moisture content and temperature of surrounding media (usually air) have on time-dependent deformations in structural wood elements depends upon their dimensions, whether they are covered with impervious materials on one or more faces, and the rate of change in temperature of surrounding media.

In summary: representations of mechanical properties of structural wood are simplistic with predicted stress or deformation based on presumption of a linear-elastic response. Predicted stress and deformation under design loads are compared with limiting values acceptable to regulatory bodies. It is the limiting values (not predicted stress and deformation) that reflect any deviations from reference conditions, e.g. effect of non-standard moisture conditions, effect of sustained load. Although limiting values are derived and design regulations stated in various ways, account is always taken of the need to avoid, at an acceptable level of risk, various undesirable results (limit states) for structural elements and systems. Limit states may be associated with global or local failure under the effect of exceptional loads, or unserviceable performance under everyday loads. Limiting values of stress (design properties), or deformation, are based on principles of mechanics and materials science considerations, and past experience. The critical requirement is that assumptions are consistent between design level analysis, procedures for assigning design properties, and definition of design loads.

Assignment of design properties for structural wood is beyond the scope of this book. Many other texts deal with that subject in detail, and it is suggested that interested readers start with USDA (1989), Madsen (1992) and Blass *et al.* (1995).

3.6   References

Alkan, S. (1986) 'Variation of some properties of a red pine', MScFE thesis, University of New Brunswick, Fredericton, NB, Canada.

Andriamaitantosa, L.D. (1995) 'Creep', STEP lecture A19. In: Timber Engineering Step 1: Basis of design, material properties, structural components and joints, Eds.Blass, H.J., Aune, P., Choo, B.S., Gorlacher, D.R., Hilso, B.O., Racher, P. and Steck, G., Centrum Hout, Almere, The Netherlands: A19/1–A19/5.

Armstrong, L.D. and Kingston, R.S.T. (1962) 'The effect of moisture content changes on the deformation of wood under stress', *Australian Journal of Applied Science*, **13**(4): 257–276.

Bach, L. (1965) 'Non-linear mechanical behavior of wood in longitudinal tension', PhD thesis, State University College of Forestry, Syracuse University, Syracuse, NY, USA.

Blass, H.J., Aune, P., Choo, B.S., Gorlacher, R., Griffith, D.R., Hilson, B.O., Racher, P. and Steck, G. (1995) Timber Engineering Step 1: Basis of design, material properties, structural components and joints, Centrum Hout, Almere, The Netherlands: A19/1–A19/5.

Bodig, J. and Goodman, J.R. (1973) 'Prediction of elastic parameters of wood', *Wood Science*, **5**(4): 249–264.

Brown, H.P., Panshin, A.J. and Forsaith, C.C. (1952) *The Physical, Mechanical and Chemical Properties of Commercial Woods of the United States*, Volume II, McGraw-Hill, New York, NY, USA.

Cave, I.D. (1969) 'The longitudinal Young's modulus of *Pinus radiata*', *Wood Science and Technology*, **3**: 40–48.

Cave, I.D. (1976) 'Modelling the structure of the softwood cell wall for computation of mechanical properties', *Wood Science and Technology*, **10**: 10–28.

Cave, I.D. (1978) 'Modelling moisture-related mechanical properties of wood. Part I: properties of the wood constituents', *Wood Science and Technology*, **12**: 75–86.

Cave, I.D. and Walker, J.C.F. (1994) 'Stiffness of wood in fast-grown plantation softwoods: The influence of microfibril angle', *Forest Products Journal*, **44**(5): 43–48.

Chen, W.F. and Han, D.J. (1988) Plasticity for Structural Engineers, Springer-Verlag. New York, NY, USA.

Chiang, Y.J. (1983) 'Design of mechanical joints in composites', PhD thesis, University of Wisconsin, Madison, WI, USA.

Cousins, W.J. (1976) 'Youngs modulus of hemicellulose as related to moisture content', *Wood Science and Technology*, **10**: 9–17.

Cowin, S.C. (1979) 'On the strength anisotropy of bone and wood', *ASME Journal of Applied Mechanics*, **46**(4): 832–837.

Dinwoodie, J.M., Higgins, J.A., Paxton, B.H. and Robson, D.J. (1991) 'Quantifying, predicting and understanding the mechanism of creep in board materials', *Proceedings of COST 508 Workshop on Fundamental Aspects of Creep in Wood*, Office of Publications of the European Communities, Luxembourg: 98–118.

Evans, R. and Ilic, J. (2001) 'Rapid prediction of wood stiffness from microfibril angle and density', *Forest Products Journal*, **51**(3): 53–57.

Evans R., Ilic, J. and Matheson, C. (2000) 'Rapid estimation of solid wood stiffness using SilviScan', *Proceedings of 26th Forest Products Research Conference*, June 19–21 2000, CSIRO, Clayton, Victoria, Australia: 49–50.

Francois, P. (1992) 'Plasticité du bois en compression multiaxiale, application à l'absorption d'énergie mécanique', Doctoral Thesis, l'université Bordeaux T, Bordeaux, France.

Gibson, L.J. and Ashby, M.F. (1988) Cellular Solids: Structure and Properties, Pergamon Press, Oxford, UK.

Gillis, P. (1972) 'Orthotropic elastic constants of wood', *Wood Science and Technology*, **6**: 138–156.

Gong, M. and Smith, I. (2000) 'Failure of softwood under static compression parallel to grain', *Journal of Institute of Wood Science*, **15**(4, issue 88): 204–210.

Gong, M. and Smith, I. (2003) 'Effect of waveform and loading sequence on low-cycle compressive fatigue life of spruce', *ASCE Journal of Materials in Civil Engineering*, **15**(1): 93–99.

Gotoh, M. (1977) 'A theory of plastic anisotropy based on a yield function of fourth order (plane stress state)', *International Journal of Mechanics and Science*, **19**: 505–520.

Hashin, Z., Bagchi, D. and Rosen, W. (1974) 'Non-linear behavior of fiber composite laminates', Contractor Report 2313, NASA, Washington, DC.

Hill, R. (1947) 'A theory of the yielding and plastic flow of anisotropic metals', *Proceedings of the Royal Society, Series A*, **193**: 281–297.

Hill, R. (1950) The Mathematical Theory of Plasticity, Oxford University Press, Oxford, UK.

Hillis, W.E. and Rozsa, A.N. (1985) 'High temperature chemical effects on wood stability. Part 3: The effect of heat on rigidity and the stability of radiata pine', *Wood Science and Technology*, **19**: 93–102.

Johansen, K W. (1949) 'Theory of timber connectors', Publication No. 9. International Association for Bridge and Structural Engineering, Bern, Switzerland: 249–262.

Kahle, E. and Woodhouse, J. (1994) 'The influence of cell geometry on the elastic constants of softwood', *Journal of Materials Science*, **29**: 1250–1259.

Kharouf, N. (2001) 'Post-elastic behavior of bolted connections in wood', PhD thesis, McGill University, Montreal, QC, Canada.

Koponen, S., Toratti, T. and Kanerva, P. (1989) 'Modelling longitudinal elastic and shrinkage properties of wood', *Wood Science and Technology*, **23**: 55–63.

Koponen, S., Toratti, T. and Kanerva, P. (1991) 'Modelling elastic and shrinkage properties of wood based on cell structure', *Wood Science and Technology*, **25**: 25–32.

Kaya, F. and Smith, I. (1993) 'Variation in crushing strength and some related properties of a red pine', *Wood Science and Technology*, **27**: 229–239.

Larsen, H.J. (2003) 'Design of structures based on glulam, LVL and other solid timber products', In: Timber Engineering, Ed. S. Thelandersson and H.J. Larsen, John Wiley & Sons, Chichester, UK.

Liu, J.Y. (1984) 'Evaluation of the tensor polynomial strength theory of wood', *Journal of Composite Materials*, **18**: 216–226.

Louranco, P.B., De Borst, R. and Rots, J.G. (1997) 'A plane stress softening plasticity model for orthotropic materials', *International Journal for Numerical Methods in Engineering*, **40**: 4033–4507.

MacKay, G.D.M. (1967) 'Mechanism of thermal degradation on cellulose', Publication 1201, Department of Forestry, Ottawa, ON, Canada.

Madsen, B. (1992) 'Structural behaviour of timber', Timber Engineering Ltd, North Vancouver, BC, Canada.

Mark, R.E. (1967) Cell Wall Mechanics of Tracheids, Yale University Press, New Haven, CT, USA.

Megraw, R., Bremer, D., Leaf, G. and Roers, J. (1999) 'Stiffness of loblolly pine as a function of ring position and height, and its relationship to microfibril angle and specific gravity', *Proceedings of 3rd Workshop Connection Between Silviculture and Wood Quality Through Modeling Approaches*. IUFRO working party S5.01–04, La Londles Maures, France, International Union of Forestry Research Organizations, Vienna, Austria: 341–349.

Moses, D. (2000) 'Constitutive and analytical models for structural composite lumber as applied to bolted connections', PhD thesis, University of British Columbia, Vancouver, BC, Canada.

Mullins, E.J. and McKnight, T.S. (1981) Canadian Woods: Their properties and uses, University of Toronto Press, Toronto, ON, Canada.

Nielsen, L.F. (1997) 'On strength of porous material', Report R-26, Department of Structural Engineering and Materials, Technical University of Denmark, Lyngby, Denmark.

Norris, C.B. (1962) 'Strength of orthotropic materials subjected to combined stresses', Report 1816, Forest Products Laboratory, Madison, WI, USA.

Patton-Mallory, M., Pellicane, P.J. and Smith, I.W. (1997) 'Nonlinear material models for analysis of bolted wood connections', *ASCE Journal of Structural Engineering*, **123** (8): 1063–1070.

Persson, K. (1997) 'Modelling of wood properties by a micromechanical approach', Report TVSM-3020, Division of Structural Mechanics, University of Lund, Lund, Sweden.

Petit, P.H. and Waddoups, M.E. (1969) 'A method of predicting the nonlinear behavior of laminated composites', *Journal of Composite Materials*, **3**: 2–19.

Preston, R.D. (1974) The Physical Biology of Plant Cell Walls, Chapman & Hall, USA.

Price, A.T. (1928) 'A mathematical discussion on the structure of wood in relation to its elastic properties', *Philosophical Transactions of the Royal Society*, **228**: 1–62.

Rahman, M.U., Chiang, Y.J. and Rowlands, R.E. (1991) 'Stress and failure analysis of double-bolted joints in Douglas-fir and Sitka spruce', *Wood and Fiber Science*, **23**(4): 567–589.

Rouger, F., Le Govic, C., Crubile, P., Soubret, R. and Paquet, J. (1990) 'Creep behaviour of French wood', *Proceedings of International Timber Engineering Conference*, Science University of Tokyo, Japan, **2**: 330–336.

Rowlands, R.E. (1985) 'Composite strength theories and their experimental correlation'. In Handbook of Composites, Eds. Kelly, A. and Robotnov, Y., North Holland, New York, NY, USA: Vol. **3**: 71–125.

Sakurada, I., Nukushina, Y. and Ito, T. (1962) 'Experimental determination of the elastic modulus of the crystalline region of oriented polymers', *Journal of Polymer Science*, **57**: 651–660.

Sandhu, R.S. (1976) 'Non-linear behavior of unidirectional and angle ply laminates', *Journal of Aircraft*, **13**(2): 104–111.

Shih, C.F. and Lee, D. (1978) 'Further developments in anisotropic plasticity', *ASME Journal of Engineering Materials and Technology*, **100**: 294–302.

Smith, I., Alkan, S. and Chui, Y.H. (1991) 'Variation of dynamic properties and static bending strength of a plantation-grown red pine', *Journal of the Institute of Wood Science*, **12**(4): 221–224.

Stefansson, F. (1995) 'Mechanical properties of wood at microstructural level', Masters thesis, Report TVSM-5057, Division of Structural Mechanics, Lund University, Lund, Sweden.

Tabarsa, T. and Chui, Y.H. (1996) 'Effects of hot pressing on properties of white spruce', *Forest Products Journal*, **47**(5): 71–76.

Tan, D. and Smith, I. (1999) 'Failure in-the-row model for bolted timber connections', *ASCE Journal of Structural Engineering*, **125**(7): 713–718.

Tsai, S.W. and Wu, E.M. (1971) 'A general theory of strength of anisotropic materials', *Journal of Composite Materials*, **5**: 58–80.

USDA (United States Department of Agriculture). (1989) Handbook of Wood and Wood-based Materials for Engineers, Architects and Builders, Hemisphere Publishing, New York, NY, USA.

Valliappan, S., Boonlaulohr, P. and Lee, I.K. (1976) 'Non-linear analysis for anisotropic materials', *Journal of Composite Materials*, **5**: 58–80.

van der Put, T.A.C.M. (1987a) 'Derivations of a general rheologic model based on the theory of molecular deformation kinetics', Report 25-87-58/24-HA-37, Faculty of Civil Engineering, Technical University of Delft, The Netherlands.

van der Put, T.A.C.M. (1987b) 'Theoretical derivations of the phenomenological laws of wood with the aid on the theory of molecular deformation kinetics', Report 25-87-58/24-HA-39, Faculty of Civil Engineering, Technical University of Delft, The Netherlands.

Vaziri, R., Olsen, M.D. and Anderson, D.L. (1993) 'Finite element analysis of fibrous composite structures: a plasticity approach', *Computers and Structures*, **44**: 103–116.

Whale, L.R.J. (1988) 'Deformation characteristics of nailed or bolted timber joints subjected to irregular short or medium term lateral loading', PhD thesis, Polytechnic of the South Bank, London, UK.

Whang, B. (1969) 'Elasto-plastic plates and shells', *Proceedings of Symposium on Application of Finite Element Method in Civil Engineering*, Vanderbilt University, Nashville, Tennessee, USA: 481–515.

Appendix: Notation

(primary or recurring items only)

A_{ijkl} = fourth order tensor of anisotropic strength parameters

$c_{H(m)}$ = stiffness adjustment factor for moisture, for hemicellulose

$c_{L(m)}$ = stiffness adjustment factor for moisture, for lignin

C = elastic compliance matrix

D = elastic stiffness matrix

D^{ep} = elasto-plastic stiffness matrix

D_H = elastic stiffness matrix for hemicellulose

D_L = elastic stiffness matrix for lignin

$\hat{D}^{S_2}$ = stiffness matrix in cell wall co-ordinates

D^{S_2} = stiffness matrix in microfibril co-ordinates

D_{LL} = stiffness of one, or a clump of, cells in L direction

D_{LL}^{cw} = stiffness of the cell wall in L direction

$\hat{D}_{LL}^{S_2}$ = stiffness of the S_2 layer in the longitudinal cell axis direction

E_i = elastic constant (elastic modulus/spring constant)

$E_{L,dyn}$ = dynamic elastic modulus for L direction

f = yield function

$F_i F_{ij}$ = strength tensor

FSP = fibre saturation point

G = transformation matrix

G_{ij} = shear modulus

H' = plastic modulus (effective stress vs. effective strain)

J_c = creep compliance

J_f = flow compliance

J_0 = elastic compliance

k = scalar parameter representing the reference yield stress

l_l = width of the latewood zone within a growth ring

l_r = width of growth ring

l_t = width of the transition zone within a growth ring

L = longitudinal direction (local co-ordinates)

L_o = longitudinal direction (log co-ordinates)

m = moisture content (%)

m_i = moisture content associated with strength X_i

MFA = microfibril angle

p = proportional cell wall volume

R = radial direction (local co-ordinates)

RD = relative density

R_o = radial direction (log co-ordinates)

S = shear strength

SL = stress level

S^{cw} = cell wall ratio

S_{av}^{cw} = average cell wall ratio

t = time

T = tangential direction (local co-ordinates)

T_g = glass transition temperature

T_o = tangential direction (log co-ordinates)

Tmp = temperature

X = tensile strength in x direction

X' = compressive strength in x direction

X_i = measured or predicted strength value at m_i

Y = tensile strength in y direction

Y' = compressive strength in y direction

α^i = shrinkage coefficients for i direction

α_{ij} = second order tensor describing origin of the yield surface

γ = shear strain

$\partial f / \partial \sigma$ = flow vector

ε = total strain

ε^e = elastic strain

ε^e = elastic strain

ε^m = moisture induced strain

ε^p = plastic strain

ε^{Tmp} = Temperature induced strain vector

ε_{dr} = delayed recoverable strain

ε_o = instantaneous recoverable strain

ε_v = viscous strain

λ_i = coefficient of traction

λ^i = thermal expansion coefficient for i

ν_{ij} = Poisson's ratio

ρ = density

ρ_e = density of earlywood

ρ_l = density of latewood

ρ_o = bulk density of cell wall

ρ_t = density of transition wood

σ = stress vector

σ_i = axial stress in i direction

σ_{ic} = compressive yield stress/strength in i direction

σ_{it} = tensile yield stress/strength in i direction

σ_{ij} = second order stress tensor

τ = shear stress

φ = microfibril angle for the S_2 layer

4

Principles of Fracture Mechanics

4.1 The Failure Stress-based Strength Theory—Motivation for Fracture Mechanics

Mechanics of materials theory and the concepts of stress and strain that it introduces have been the foundation of structural engineering for several hundred years. The approach is straightforward and elegant, and it has served as the basis for an encyclopedic list of analysis techniques. And yet, in many practical situations its predictions are wrong.

The most widely cited early work in fracture mechanics was that published by A. A. Griffith in 1921 (Griffith, 1921). Among his experiments were tests on thinly drawn glass fibres. His results showed that thinner fibres failed at higher stresses than thicker fibres, an outcome not predicted by stress-based strength theory. Indeed, rupture stresses increased by an order of magnitude when fibre diameters dropped from 100 μm down to less than 10 μm. Concurring results were obtained by da Vinci over 400 years prior, who tested wire rope of different lengths. On a more practical side, numerous structural failures as documented by Broek (1986), among others, were caused not by a poor application of stress theory, but rather to a misunderstanding of the role of cracks in engineering materials.

At an even more fundamental level, one can consider typical coefficients of variation of yielding in steel compared to rupture of wood in tension. The former carries a typical COV of about 2–3%, while the latter carries a COV of 20–40%. One could argue these results represent examples of materials that do and do not fit the models used to describe them.

Fracture mechanics is a series of models used to describe the influence of cracks and defects on material behaviour. The work of Griffith proved that cracks and defects dictate the strength of a material more than any other single feature. In this chapter, concepts of fracture mechanics are discussed with an ultimate goal of applications to wood and wood composites.

Fracture and Fatigue in Wood   I. Smith, E. Landis and M. Gong
© 2003 John Wiley & Sons, Ltd   ISBN: 0-471-48708-2 (HB)

4.2 Griffith Theory

4.2.1 Stress concentrations

In a classic paper by Inglis (1913), a solution is presented for the stresses occurring around an elliptical hole in a plate subjected to uniaxial tension (illustrated in Figure 4.1). The greatest concentration of stress occurs at point p, and is defined by:

$$\sigma_y = \sigma \left[1 + 2\sqrt{\frac{a}{\rho}} \right] \tag{4.1}$$

where σ_y is the vertical stress at p, σ is the stress far away from the hole, a is half the major axis of the hole, and ρ is the radius at the tip. Inspection of Equation (4.1) shows that the magnitude of the stress at the tip of the hole increases both with the size of the hole and with the sharpness of the tip. The relevance of this solution for strength of materials problems is that even small cracks or other discontinuities in a material can have tremendous impact on local stresses. Indeed, a small crack with a sub-micron tip radius can amplify stresses several hundred times, causing a structural member to fail by fracture under relatively low average stress.

4.2.2 Condition for crack growth

While the stress concentrations provided by cracks presents a mechanism for fracture at low stresses, it does not provide a quantitative criterion for crack growth. Griffith's

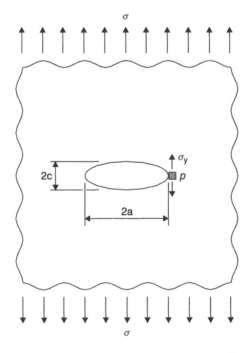

Figure 4.1   Illustration of elliptical hole used in Inglis solution.

approach (Griffith, 1921) was to consider the thermodynamic equilibrium of a crack system. If an elastic body is deformed by an external load, there is elastic strain energy stored in the body in addition to a change in the potential energy of the load system (equal to the negative work of load). The presence of a crack is accounted for in the energy balance by the surface energy of the crack surfaces. The total energy of the system, Π, may be written as the sum of these components:

$$\Pi = U + U_P + W$$
$$= U - F + W \tag{4.2}$$

where U is the strain energy, U_P is the potential energy of the load system, F is the external work of load, and W is the surface energy associated with crack formation. According to Griffith, if the crack is to grow, the total energy must be either reduced or unchanged. This may be written in terms of system equilibrium:

$$\frac{d\Pi}{dA} = \frac{d}{dA}(U - F + W) = 0 \tag{4.3}$$

or

$$\frac{d}{dA}(F - U) = \frac{dW}{dA} \tag{4.4}$$

where dA is the incremental change in crack area. Equation (4.4) represents the Griffith criterion for crack growth, which may be illustrated by a consideration of the cracked body shown Figure 4.2. In this case, the crack may be represented by an ellipse with a minor axis that approaches zero (dimension c in Figure 4.1). Then the Inglis's solution can be applied to evaluate the strain energy:

$$U = \frac{\pi \sigma^2 a^2 b}{E} \tag{4.5}$$

where b is the thickness of the plate and E is the material's modulus of elasticity.[1] For an elastic body with a crack that grows under a state of constant stress, the external work of load can be shown to be twice the internal strain energy. Specifically:

$$F = 2U \tag{4.6}$$

The surface energy of the crack can be written in terms of the free unit surface energy of the material, γ:

$$W = 4ab\gamma \tag{4.7}$$

where the $4ba$ term is based on two surfaces of length $2a$.

The terms of Equation (4.2) can be plotted as shown in Figure 4.3. The plot shows that the total system energy is in equilibrium at a crack length of $a = a_0$, where the slope of Π is zero. However, this is an *unstable* equilibrium. If the system is perturbed

---

[1] It should be noted that this is the plane stress solution. For plane strain the solution should be multiplied by $1 - \nu^2$, where ν is the Poisson's ratio. See any textbook on elasticity theory for explanation of plane stress and plane strain.

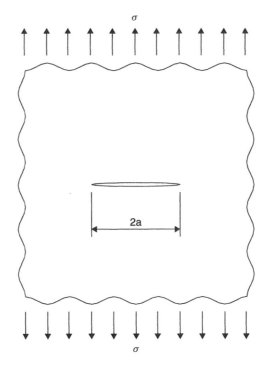

Figure 4.2   Crack in a uniform tension field.

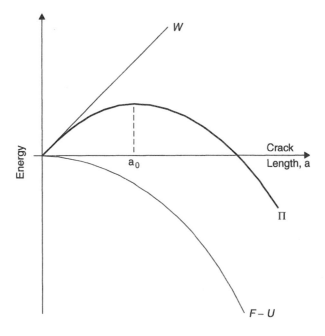

Figure 4.3   Energetics of crack growth.

such that $a > a_0$ the system will seek a lower energy state, and because Π drops indefinitely, the crack will grow without limit.

The equilibrium state can be determined by substituting Equations (4.5)–(4.7) into (4.4), noting that $\mathrm{d}A = b\,\mathrm{d}a$:

$$\frac{2\pi\sigma^2 ab}{E} = 4b\gamma \tag{4.8}$$

Solving for stress yields:

$$\sigma_f = \sqrt{\frac{2E\gamma}{\pi a}} \tag{4.9}$$

where σ_f denotes the fracture stress or fracture strength of the material. This equation clearly indicates the relationship between maximum stress and crack length. The fracture strength decreases with one over the square root of crack size. This equation can alternatively be used to determine the allowable crack size for a given stress. It should be noted that not all crack configurations are unstable. Crack stability is dictated by a number of factors, and is discussed in more detail below.

Griffith tested this theory on blown glass and drawn glass fibres, and found good agreement. It should be recognised, however, that glass is not necessarily a typical structural material. Its rather uniform homogeneous character separates it considerably from heterogeneous, anisotropic materials such as wood and wood composites. What follows are expansions and additions to Griffith theory that allow applications to a much wider range of problems.

4.3 Linear Elastic Fracture Mechanics (LEFM)

The work of Griffith formed the fundamental basis for modern fracture theory. Over the last 75 years the work has been continually expanded for application to a wide variety of problems. Two fundamental approaches to fracture problems have arisen: the strain energy release rate, which is based on the global energy balance of Griffith; and the stress intensity factor, which is based on the local stress distribution around a crack tip. An overview of these two approaches is presented in the following sections. An assumption that applies throughout this section is that the material under consideration exhibits linear or very nearly linear elastic behaviour right up to the point where fracture occurs.

4.3.1 Strain energy release rate

Referring to the energy balance and equilibrium Equations (4.2)–(4.4), the left-hand side of Equation (4.4) is commonly referred to as the strain energy release rate G, while the right-hand side of Equation (4.4) is commonly referred to as the crack resistance R. Equation (4.4) can then be written as:

$$G = \frac{\mathrm{d}}{\mathrm{d}A}(F - U) \tag{4.10}$$

and:

$$R = \frac{\mathrm{d}W}{\mathrm{d}A} \tag{4.11}$$

Thus, the critical condition for crack growth of Equation (4.4) may be rewritten as:

$$G = R \tag{4.12}$$

G and R have units of energy per unit area (e.g. J/m^2). Therefore, G is often interpreted as the energy available to grow a crack of unit area, while R is interpreted as the energy required for propagation of a crack of unit area. Frequently, the crack growth condition is written in terms of a critical strain energy release rate G_C. The condition for crack growth in this case is:

$$G = G_C \tag{4.13}$$

where G_C is often referred to as the fracture energy of a material. It should be noted that although G_C and R are used interchangeably here, G_C is most commonly used in situations where crack resistance is a material constant. As is discussed in Section 4.4.2, in many cases the resistance to crack growth in a material is not a constant, but varies with crack length, and is therefore a function of a. In these cases it is more common to use R to denote crack resistance.

Energy release rate G is determined for an arbitrary structure using Equation (4.10). As an example, consider the centre-cracked plate in a uniform tension field (Figure 4.2). Using Equations (4.5) and (4.6):

$$G = \frac{2\pi\sigma^2 ab}{E} \tag{4.14}$$

As defined above, σ is the far field stress, E is the elastic modulus, b is the thickness, and $2a$ is the crack length. Griffith's work with glass showed that the material's specific surface energy, γ, was a reasonable approximation for G_C, and that result was used to calculate the fracture strength of the material. However, for nearly all polycrystalline and polymeric materials, this will lead to a gross underestimate of the material's resistance to fracture due to numerous other microstructural mechanisms involved with crack growth. Thus, G_C must be measured for most all structural materials.

G can be calculated for an elastic body subjected to a load that may be classified as either a load or a displacement control loading condition. Figure 4.4 illustrates these two conditions and their respective load-displacement curves for a crack of length a, that has grown an amount $\mathrm{d}a$ in a specimen of thickness b. In load control, a known load is applied to the structure, and the corresponding displacement is measured. Conversely, in displacement control, the structure is forced to undergo a prescribed displacement, and the corresponding load is measured.

In the case of load control (Figure 4.4c), the strain energy in the body at crack length a may be given by:

$$U_1 = \tfrac{1}{2}P\delta \tag{4.15}$$

For the same body with crack length $a + \mathrm{d}a$, the strain energy becomes:

$$U_2 = \tfrac{1}{2}P(\delta + \mathrm{d}\delta) \tag{4.16}$$

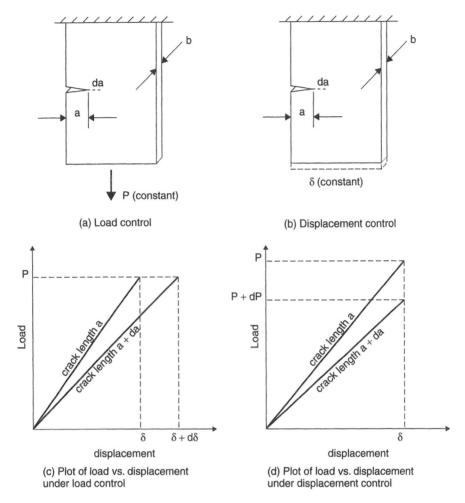

Figure 4.4 Crack growth under load control and displacement control.

So the incremental change in strain energy during this time is:

$$dU = U_2 - U_1 = \tfrac{1}{2}P(\delta + d\delta) - \tfrac{1}{2}P\delta = \tfrac{1}{2}P d\delta \qquad (4.17)$$

The incremental work, dF, done by the load during this crack growth is simply $P\,d\delta$. Substituting this result and those of Equation (4.17) into Equation (4.10) yields:

$$G = \frac{1}{b}\left[\frac{dF}{da} - \frac{dU}{da}\right] = \frac{1}{b}\left[P\frac{d\delta}{da} - \frac{1}{2}P\frac{d\delta}{da}\right] = \frac{1}{2b}P\frac{d\delta}{da} \qquad (4.18)$$

A similar analysis in displacement control yields:

$$G = -\frac{1}{2b}\delta\frac{dP}{da} \qquad (4.19)$$

It may appear that the value of G is dependent on whether the loading condition is displacement or load controlled, by virtue of the different signs in the two results. However, it can be shown that the results are in fact identical by writing Equations (4.18) and (4.19) in terms of the change in specimen compliance that accompanies crack growth. Specifically, if compliance C is defined as the reciprocal of the slope of the load-displacement curve, i.e. $C = \delta/P$, then G becomes:

$$G = \frac{1}{2b}P^2\frac{dC}{da} \tag{4.20}$$

which is valid for both load control and displacement control.

A simple and common application of the compliance Equation (4.20) is the double cantilever beam specimen used to measure fracture energy of a variety of materials. Such a specimen is shown in Figure 4.5. The specimen gets its name because it can be approximated as a pair of opposing cantilever beams. From basic mechanics of materials theory the relationship between the load and the deflection at the end of the cantilever of length L is given by:

$$\Delta = \frac{PL^3}{3EI} \tag{4.21}$$

where I is the moment of inertia for the beam, and E is the modulus of elasticity. Substituting the appropriate terms for I ($I = \frac{1}{12}bh^3$ for the rectangular cross-section shown), and substituting the crack length a for cantilever length L, and recognising that for a double cantilever the deflection is twice that of the single cantilever ($\delta = 2\Delta$) then Equation (4.21) becomes:

$$\delta = \frac{8Pa^3}{Ebh^3} \tag{4.22}$$

Thus, the compliance of the system is:

$$C = \frac{8a^3}{Ebh^3} \tag{4.23}$$

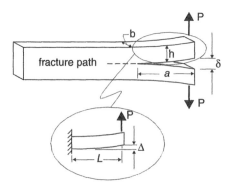

Figure 4.5   Double cantilever beam specimen.

This expression can be substituted into Equation (4.20) to produce an expression for G for this particular geometry:

$$G = \frac{1}{2b} P^2 \frac{\mathrm{d}}{\mathrm{d}a} \left(\frac{8a^3}{Ebh^3} \right) = \frac{12P^2a^2}{Eb^2h^2} \tag{4.24}$$

It should be noted that Equation (4.24) can be modified to account for material anisotropy, which is not included in the derivation of Equations (4.21)–(4.24) (Daniel and Ishai, 1994):

$$G = \frac{12P^2}{E_1b^2h} \left[\left(\frac{a^2}{h} \right) + \frac{1}{10} \left(\frac{E_1}{G_{31}} \right) \right] \tag{4.25}$$

where E_1 is the longitudinal modulus, and G_{31} is the transverse shear modulus in a unidirectional composite material.

Looking at either Equation (4.24) or (4.25), it can be seen that G increases with both the load and the crack length, meaning the energy available to propagate a crack increases with these quantities. In fracture tests, typically the load is increased (hence G increases) until the crack grows. At that point $G = G_C$.

Figure 4.6 shows a typical result from tests by Whittaker (2002). In these tests, eastern hemlock specimens 400 mm long, with an initial notch length of 70 mm were loaded perpendicular to grain (*RL* direction).[2] The tests were run in displacement control so the crack opens at a prescribed rate (more on this below). Crack length and load are plotted as a function of time in Figure 4.6a. Clearly, the load begins to fall at the time the crack begins to grow in length. During the test the strain energy release rate grows with the load until it levels off (more or less) at a G of about 170 J/m^2. The conclusion is that G_C for this material is around 170 J/m^2.

Assuming G_C can be considered a material parameter, it can be used to estimate the load carrying capacity of a structure with cracks. In this example, because G cannot exceed G_C in the material, the load capacity of the specimen must fall as the crack length increases.

The stability of the crack can be evaluated by considering what happens to the system energy as the load or displacement increases. Figure 4.7a shows plots of G (using Equation (4.24)) as a function of crack length, a, for different load levels. Clearly, in all cases, as the crack length increases, G increases. That is, more energy becomes available to propagate the crack as the crack extends. For the experiments just described (initial crack length of 70 mm), the curves indicate that the crack will propagate at a load of 450 N. Furthermore, because G increases with crack length under constant load and it does not fall below G_C, the crack will continue to propagate until the specimen is completely fractured. This represents unstable crack growth. As soon as $G = G_C$, the crack propagates indefinitely. However, in Figure 4.7b, the same experimental configuration is shown under displacement control rather than load control. Note that, in this case, G decreases as the crack length increases. In other words, there is less energy available to propagate the crack as the crack extends. This

---

[2] In this notation, L = longitudinal, R = radial, and T = tangential. The first letter indicates the direction perpendicular to the crack plane and the second letter indicates the direction of crack growth. Thus, an *RL* crack propagates along the tree axis in a plane that is tangential to the growth rings, i.e. the plane is perpendicular to the radial direction. See Figure 5.1 for an illustration.

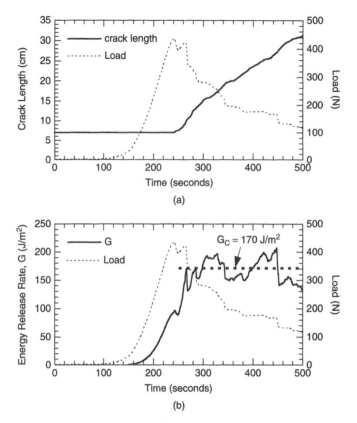

Figure 4.6 Plots of load, crack length, and energy release rate for a double cantilever beam of eastern hemlock.

is due to a relaxation of the specimen. Now, for the same experimental conditions, it is again seen that the crack will propagate when $G = G_C$ (at a displacement of 3.4 mm). However, when the crack length grows, G falls below G_C, and will not extend further until the displacement is increased such that $G = G_C$ again. This represents stable crack growth.

Clearly, there are experimental advantages to testing specimens in a configuration that produces stable crack growth. Variations in G_C can be evaluated in a single specimen along a crack face. Whereas in unstable crack configurations, all that can be measured is the energy to initiate crack growth.

To summarise, the strain energy release rate G, is a fracture criteria based on the energy balance work of Griffith. If the critical strain energy release rate for a material is known, the fracture strength of the structure can be predicted.

4.3.2   The stress intensity factor

The traditional problem with the strain energy release rate as a measure fracture energy and fracture toughness is that it relies on a global assessment of the energy state of a structure before and after crack growth. This is often not practical for large systems,

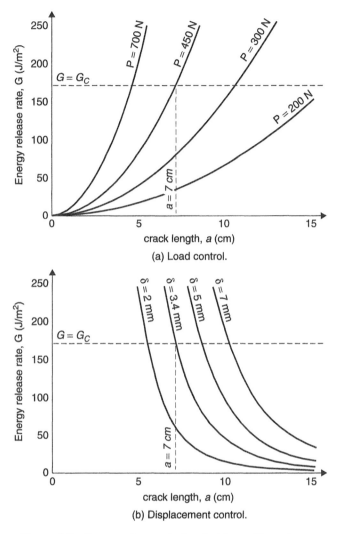

Figure 4.7 Energy release rate for double cantilever beam.

and the method can quickly become intractable. As an alternative, a fracture criterion has been developed that is based on the local stress state at the crack tip.

Consider the tip of a crack in a uniform tension field, σ, as shown in Figure 4.8. In two dimensions, the general solution for the stresses near the tip of the crack (where the origin of the reference coordinates is at the crack tip) is as follows (Anderson, 1995):

$$\sigma_x = \frac{K_{\mathrm{I}}}{\sqrt{2\pi r}} \cos \frac{\theta}{2} \left(1 - \sin \frac{\theta}{2} \sin \frac{3\theta}{2}\right)$$

$$\sigma_y = \frac{K_{\mathrm{I}}}{\sqrt{2\pi r}} \cos \frac{\theta}{2} \left(1 + \sin \frac{\theta}{2} \sin \frac{3\theta}{2}\right) \qquad (4.26)$$

$$\tau_{xy} = \frac{K_{\mathrm{I}}}{\sqrt{2\pi r}} \sin \frac{\theta}{2} \cos \frac{\theta}{2} \cos \frac{3\theta}{2}$$

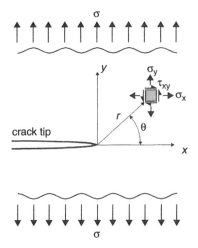

Figure 4.8   Stresses at a crack tip.

The constant K_I is known as the stress intensity factor, and is defined by:

$$K_I = \lim_{r \to 0} \sqrt{2\pi r}\, \sigma_y \bigg|_{\theta=0} \tag{4.27}$$

K_I represents the strength of the stress singularity that occurs at the tip of a sharp crack. It is a function of the specimen geometry, applied load, and crack length. It can thus be defined from dimensional analysis to be:

$$K_I = \beta \sigma \sqrt{a} \tag{4.28}$$

where β is a dimensionless geometry parameter, σ is the far field stress, and a is the crack length. The units of K_I are stress times the square root of length (e.g. $N \cdot m^{-3/2}$).

Because it defines the intensity of the stress at the crack tip, K_I can serve as a fracture toughness parameter. That is, a crack in a material will grow when K_I reaches a critical value:

$$K_I = K_{IC} \tag{4.29}$$

where K_{IC} is the critical stress intensity factor for a material. K_{IC} is considered a material property (analogous to yield strength) that defines the material's resistance to crack growth. It is often referred to as the fracture toughness of the material. It should be noted that the I subscript indicates mode I fracture; mode I being in-plane tension. Fracture modes II (in-plane shear) and III (out-of-plane shear) are illustrated in Figure 4.9. Each of these fracture modes has independent stress intensity factors, K_{II} and K_{III}, respectively. Mode I is typically the dominant case for most structural materials, and as such generally receives the most attention in the literature. However for wood, mode II and mode III can be dominant depending on the relationship between the principal stress axes and the material axes. A discussion of mixed mode fracture is presented in Section 4.3.4.

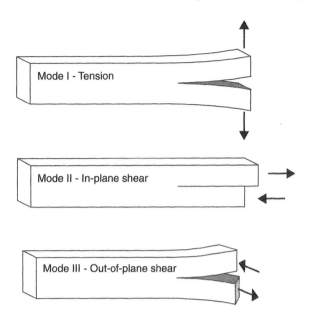

Figure 4.9 Illustration of the three fracture modes.

Practical applications of stress intensity factors are generally accomplished using the wide range of published values for different geometry. Table 4.1 has tabulated stress intensity factors for some useful geometries. For example, consider the centre-cracked plate whose crack is small compared to its width. The mode I stress intensity factor for this configuration is given by:

$$K_I = \sigma \sqrt{\pi a} \tag{4.30}$$

(Thus the geometry factor, β, in this case is equal to $\sqrt{\pi}$.) To determine the stress at which the crack will grow, the critical stress intensity factor is substituted into Equation (4.30) to yield:

$$\sigma_f = \frac{K_{IC}}{\sqrt{\pi a}} \tag{4.31}$$

where σ_f denotes the fracture stress.

As an alternative example, if a notched tension specimen of red spruce is loaded perpendicular to the grain (*TL* direction), as shown in Figure 4.10, we can calculate the ultimate load based on tabulated values of critical stress intensity factors for the species. From Table 4.1, the mode I stress intensity factor is:

$$K_I = 1.12\sigma \sqrt{\pi a} \tag{4.32}$$

(Note this assumes a small a/w ratio. Carrying out the longer expression for β in Table 4.1 one can show this to be valid.) From the *Wood Handbook* (USDA, 1999), the fracture toughness for red spruce in the *TL* direction is listed as 420 kN·$m^{-3/2}$. Solving Equation (4.32) for stress, and substituting the value for K_{IC} in for K_I, produces

Table 4.1   Stress intensity factors for common geometries

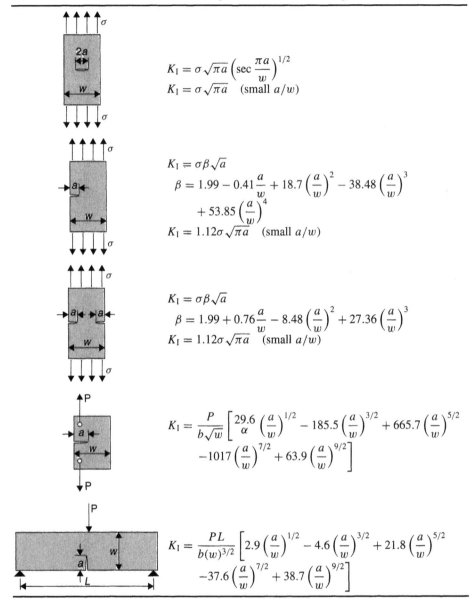

$$K_I = \sigma\sqrt{\pi a}\left(\sec\frac{\pi a}{w}\right)^{1/2}$$
$$K_I = \sigma\sqrt{\pi a} \quad (\text{small } a/w)$$

$$K_I = \sigma\beta\sqrt{a}$$
$$\beta = 1.99 - 0.41\frac{a}{w} + 18.7\left(\frac{a}{w}\right)^2 - 38.48\left(\frac{a}{w}\right)^3$$
$$+ 53.85\left(\frac{a}{w}\right)^4$$
$$K_I = 1.12\sigma\sqrt{\pi a} \quad (\text{small } a/w)$$

$$K_I = \sigma\beta\sqrt{a}$$
$$\beta = 1.99 + 0.76\frac{a}{w} - 8.48\left(\frac{a}{w}\right)^2 + 27.36\left(\frac{a}{w}\right)^3$$
$$K_I = 1.12\sigma\sqrt{\pi a} \quad (\text{small } a/w)$$

$$K_I = \frac{P}{b\sqrt{w}}\left[\frac{29.6}{\alpha}\left(\frac{a}{w}\right)^{1/2} - 185.5\left(\frac{a}{w}\right)^{3/2} + 665.7\left(\frac{a}{w}\right)^{5/2}\right.$$
$$\left. -1017\left(\frac{a}{w}\right)^{7/2} + 63.9\left(\frac{a}{w}\right)^{9/2}\right]$$

$$K_I = \frac{PL}{b(w)^{3/2}}\left[2.9\left(\frac{a}{w}\right)^{1/2} - 4.6\left(\frac{a}{w}\right)^{3/2} + 21.8\left(\frac{a}{w}\right)^{5/2}\right.$$
$$\left. -37.6\left(\frac{a}{w}\right)^{7/2} + 38.7\left(\frac{a}{w}\right)^{9/2}\right]$$

the stress in the material when the cracks start to propagate and the specimen fails:

$$\sigma_f = \frac{K_{IC}}{1.12\sqrt{\pi a}} = \frac{(420 \text{ kPa}\sqrt{\text{m}})}{1.12\sqrt{\pi(0.01 \text{ m})}} = 2.1 \text{ MPa}$$

An alternative interpretation of this example is in the relationship between the crack size and the fracture strength. Often, for a particular structure under a given stress, it is desirable to work in terms of a 'critical crack length' a_c in assessing structural safety.

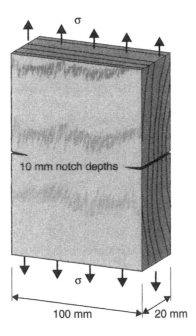

Figure 4.10 Notched tension specimen.

The critical crack length is the crack size in a structure at incipient fracture. It can be evaluated by solving for a in Equation (4.31):

$$a_c = \frac{K_{\mathrm{IC}}^2}{\pi \sigma_f^2} \qquad (4.33)$$

The critical crack length is often used in conjunction with periodic structural inspections, particularly in the aerospace industry, or as part of quality control during manufacturing.

4.3.3 Reconciliation of K and G

As is demonstrated in the preceding sections, both K and G can be used to predict the fracture strength of a particular structure even though the development of these two fracture parameters was quite different. The strain energy release rate was developed using a global energy balance, while the stress intensity factor was developed evaluating the stress field at the tip of a crack. Needless to say that, if they are indeed accurate predictors, there should be some way to reconcile the two parameters.

A simple way to reconcile K and G is to look at their values for a common geometry. For the centre-cracked plate of unit thickness, it is known from the above that:

$$K_{\mathrm{I}} = \sigma\sqrt{\pi a} \quad \text{and } G_{\mathrm{I}} = \frac{\pi\sigma^2 a}{E}$$

If both equations are solved for σ, and set equal to each other, then:

$$G_I = \frac{K_I^2}{E} \text{ (plane stress)} \tag{4.34}$$

$$G_I = (1 - \nu^2)\frac{K_I^2}{E} \text{ (plane strain)} \tag{4.35}$$

The advantages and disadvantages of each parameter in practical applications should be apparent. As discussed above, the disadvantage of G is its global nature. A relationship between work of load and strain energy must be known for an impending crack growth. The advantage of this approach, however, is that it is not as sensitive to the precise crack tip geometry. In the case of the stress intensity factor K, a well-defined crack tip geometry is necessary for proper evaluation of the crack tip stress field. However, it is not necessary to evaluate the influence of the individual crack growth on the entire structure.

It may be argued that G is a more fundamental quantity in that it is derived directly from first principles, and its units have direct physical meaning: the energy required to create new crack surfaces. K on the other hand is an integration constant whose units do not have a direct physical meaning.

4.3.4 Mixed mode fracture

While the discussions of fracture thus far have been confined to 'pure' fracture modes I, II and III, cracks in real structures tend to be subjected to combinations of fracture modes, and must be considered as 'mixed mode'. This can occur in a crack that is inclined relative to the axis of principal stress, or in a crack that is subjected to a multi-axial stress state. (Noting, of course, that these two cases are equivalent through the laws of stress transformation.)

Although crack tip stresses under mixed mode loading can be added through linear superposition, stress intensity factors of different modes cannot. Typically, each mode has its own independent critical stress intensity factor: K_{IC}, K_{IIC} and K_{IIIC}, and crack propagation occurs when any one of these is reached. Strain energy release rate, being a scalar quantity, can be added in mixed mode conditions: $G = G_I + G_{II} + G_{III}$. It should be noted, however, that traditional solutions for stress intensity factors and strain energy release rates assume that a planar crack will extend along the plane of the existing crack. This is not what is typically observed.

If an angled crack in a uniform tensile field as shown in Figure 4.11 is considered, the general observation is that angled cracks will turn towards the plane perpendicular to the principal tensile stress as shown in the figure. The reason for this can be shown quantitatively, but is also fairly intuitive: the crack will grow in the direction that maximises the energy release. By evaluating the mode I and mode II stress intensity factors at the crack tip for a range of crack extension angles, it can be shown that G is maximised when the crack extension runs perpendicular to σ (Williams and Ewing, 1972).

It must be emphasised that the direction of crack extension is governed not only by the direction that maximises G, but also by the natural planes of weakness in the material. This is especially true for wood where differences in fracture toughness relative to the grain direction are enormous. Crack propagation direction is therefore dictated

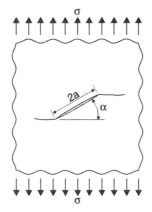

Figure 4.11   Inclined crack in a uniform tension field.

by a combination of direction of maximum energy release rate G and minimum crack resistance R. Thus cracks will typically propagate parallel to the grain (minimum R in the material), but will frequently jump between grain lines when doing so maximises energy release.

4.3.5   Fatigue in a fracture mechanics context

Often fracture of materials is considered at a limit state, such as the ultimate strength. Fracture theory says that cracks will grow when the crack tip stress intensity factor reaches the critical value for that material. In fatigue loading, interest is in the conditions for crack growth under repetitive loading cycles at loads well below the critical value. That is, the problem becomes one of predicting crack growth under conditions in which $K_\mathrm{I} \ll K_\mathrm{IC}$. The fracture mechanics approach is to relate change in crack length to number of load cycles and the range of stress intensity factors. Such a relationship can be expressed as:

$$\frac{\mathrm{d}a}{\mathrm{d}N} = f(\Delta K) \tag{4.36}$$

where a is the crack length, N is the number of cycles at a range of stress intensity factor ΔK. $\mathrm{d}a/\mathrm{d}N$ may be thought of as the crack growth per load cycle.

Relationships such as that of Equation (4.36) can be based on empirical relationships such as the one shown in Figure 4.12. Examination of this figure reveals what would be intuitively expected; the greater the range in stress (and therefore stress intensity factor), the faster the cracks will grow from an initial crack length of a_0. Indeed, in all cases crack growth is initially rather slow, but the rate increases dramatically with the number of cycles. Clearly, the closer the initial crack length is to the critical crack length, the sooner the critical crack length will be reached for a given stress range.

The most common assumption about fatigue crack growth is that it follows a power law relationship such as:

$$\frac{\mathrm{d}a}{\mathrm{d}N} = C(\Delta K)^m \tag{4.37}$$

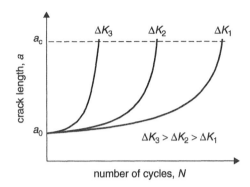

Figure 4.12   Crack growth as a function of fatigue cycles.

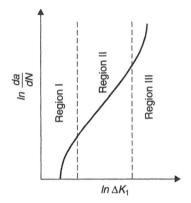

Figure 4.13   Crack growth rate as a function of stress intensity.

where C and m are empirically derived constants. This relationship is based on experimental observations such as that illustrated in Figure 4.13. Clearly, the relationship is only valid for the linear region, however it gives a basis by which crack growth can be estimated for cyclic service load conditions.

4.4   Nonlinear Fracture Mechanics

In the previous section, concepts are described that apply primarily to materials that exhibit essentially linear elastic stress-strain relationships right up to rupture. While materials such as glass and cast iron are good examples of materials for which this is appropriate, they can hardly be considered typical of most modern structural materials. The basic assumption of LEFM is that all available strain energy goes into propagating a crack, or in the view of Griffith, the creation of new material surfaces. In nearly all materials, there are numerous other microstructural mechanisms that are capable of dissipating strain energy: plastic deformation around the crack tip in metals, microcracking and friction in concrete and rock, fibre bridging in wood and fibrous composites. These and other 'toughening mechanisms' are discussed in more detail below, however, they all affect measured fracture energy to a varying degree. The degree to which the various toughening mechanisms affect fracture behaviour dictates whether LEFM can be applied to a particular material. Generally, if the effects are small, LEFM can be applied without modification. If the effects are large, then modifications to the theory must be made to account for the nonlinearity caused by the toughening mechanisms.

4.4.1 The fracture process zone

Typically, nonlinearities manifest themselves in material size effects. Although fracture toughness is considered a material property, measured fracture toughness will vary with specimen size if the nonlinearities are significant. Specifically, large specimens are observed to have lower fracture toughness than smaller specimens of the same material. The reason for this 'size effect' is the presence of a 'fracture process zone'. The fracture process zone is the region around the tip of the crack in which various toughening mechanisms are mobilised. Ahead of the crack tip plastic deformation, microcracking, or intersection with voids or interfaces might be observed, while behind the crack tip bridging by fibres, friction between crack faces, or crack branching might be observed. Clearly, the degree of heterogeneity in the material strongly influences the characteristics and extent of this region of energy dissipation.

The reason for observed size effects is that the size of a fracture process zone is essentially invariant and its relative influence tends to decrease as the specimen size increases. Thus, larger specimens exhibit behaviour closer to that predicted by LEFM. In smaller specimens the influence of the fracture process zone is greater, and they are observed to have a relatively high toughness. The limit states of these two extremes are traditional ultimate stress theory for small specimens, and LEFM for large specimens, as illustrated in Figure 4.14. Nonlinear fracture theories are aimed at bridging the gap between these two theories of failure (Bazant, 1999).

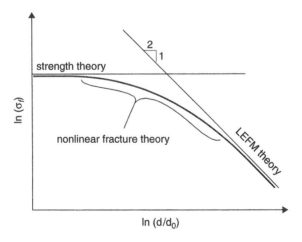

Figure 4.14 Illustration of transition from strength theory to fracture theory: d/d_0 represents the ratio of specimen size to the process zone size.

4.4.2   R-curves

A convenient tool for quantifying the influence of the fracture process zone on various fracture properties is the 'crack growth resistance curve' or *R*-curve. Crack resistance, *R*, has already been defined in Equations (4.10) and (4.11). The *R*-curve is a plot of *R* as a function of crack length, and is thus a way to represent crack resistance that is not constant. Rising *R*-curves are common among materials that exhibit toughening mechanisms described above. An example of such a curve is shown in Figure 4.15 along with some mechanisms that cause the rising *R*-curve behaviour. Essentially, rising *R*-curves reflect the fact that certain toughening mechanisms are not mobilised until the crack grows to a certain size. This is especially true of crack bridging, where bridging fibres require sufficient deformation to produce a closing force. The levelling off of the *R*-curve at longer crack lengths is an indication that the influence of the toughening mechanisms is not indefinite. At sufficiently large crack sizes, *R* becomes nominally constant as is assumed in LEFM. Such *R*-curves reinforce the notion that very large specimens approach a fracture behaviour that follows LEFM.

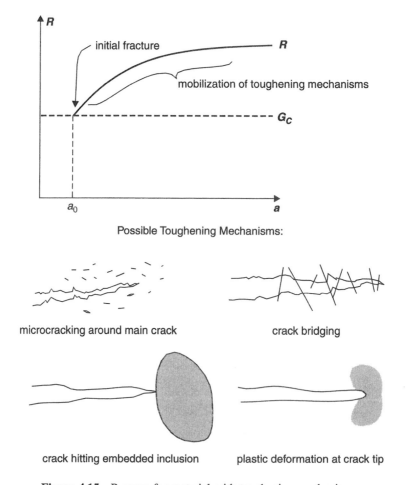

Figure 4.15   R-curve for material with toughening mechanisms.

R-curves play an important role in analysing the crack stability of different materials and different specimen geometry. As discussed in the context of LEFM, the condition for crack growth is that $G = R$ (or $G = G_C$), and crack growth can be stable or unstable depending on how G varies with crack length. Now that there is the possibility that R might not be constant, this condition gets slightly more complicated.

Consider the rising R-curve shown in Figure 4.16. On the same plot is G for a centre-cracked plate (Figure 4.2), which in Section 4.3.1 is found to be:

$$G = \frac{2\pi\sigma^2 ab}{E}$$

For this geometry G rises linearly with crack length. For a plate with length of $2a_0$ under a stress of σ_0, G is less than R; therefore the crack will not grow. If the stress in the plate is raised to σ_1, G becomes greater than R, and the crack will grow. However, if it grows under a constant stress, along the line labelled G ($\sigma = \sigma_1$), at a certain crack length (say, a_1) G will once again be less than R, and the crack will stop growing. Hence we have a condition of stable crack growth. This phenomenon will continue as the stress is increased. As long as the increase in stress causes the crack to grow such that G becomes less than R, the crack growth is stable.

Consider now what happens to the plate at crack length a_2. The crack will not grow until the stress reaches σ_3. At this point once again $G = R$ and the crack grows. But unlike the previous cases, under a constant stress of σ_3 the crack will grow indefinitely because G does not drop below R. This is clearly an unstable situation, and leads us to a general crack growth stability criterion. Looking at Figure 4.16, the stability can be seen to be dependent on the relative slopes of G and R. Thus, if

$$\frac{\mathrm{d}G}{\mathrm{d}a} < \frac{\mathrm{d}R}{\mathrm{d}a} \tag{4.38}$$

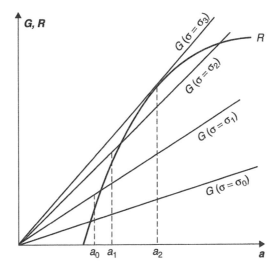

Figure 4.16   Development of crack growth stability criterion.

there is stable crack growth. Conversely, if

$$\frac{dG}{da} > \frac{dR}{da}$$

(4.39)

there is unstable crack growth. The critical situation occurs when the quantities are equal. It should be noted that this is a general condition, and it applies to both linear and nonlinear fracture theories.

4.4.3   Fictitious/effective crack models

Linear elastic fracture mechanics theory assumes a sharp crack tip for which stress fields may be calculated. While the elastic solutions predict infinite stresses at the crack tip, in real materials, yielding and damage prevent this from actually occurring. According to the discussion above, as long as the size of the fracture process zone is small LEFM theory can be applied. A number of different approaches have been used to examine the size of the process zone. These approaches have in turn been used to develop predictive models for non-linear fracture processes.

Dugdale–Barenblatt approach

Dugdale (1960) and Barenblatt (1962) separately treated a crack with a plastic zone as a slightly larger fictitious crack with closing stresses applied at the tips. Referring to Figure 4.17, the length of the plastic zone is the difference between the size of the

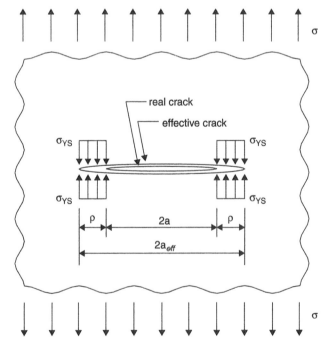

Figure 4.17   Effective crack length for plastic deformation at crack tip.

fictitious crack and the real crack, while the closing stress is equal to the yield stress of the material. This length, ρ, is determined such that the stress intensity factor goes to zero at the tip of the fictitious crack. For this to occur, the two cases (real crack subject to a far field tension stress, and fictitious crack subject to closing stresses) are superimposed. The stress intensity factor for the fictitious crack subjected to the uniform tension stress is:

$$K_I^\sigma = \sigma \sqrt{\pi(a + \rho)} \tag{4.40}$$

while the stress intensity factor due to the closing stress of σ_{YS} can be shown to be:

$$K_I^\rho = 2\sigma_{YS}\sqrt{\frac{a + \rho}{\pi}} \cos^{-1}\frac{a}{a + \rho} \tag{4.41}$$

Since the requirement was that the stress intensity factor vanishes, $K_I^\sigma = -K_I^\rho$ and the solution for ρ is:

$$\rho = \frac{\pi^2\sigma^2 a}{8\sigma_{YS}^2} = \frac{\pi K_I^2}{8\sigma_{YS}^2} \tag{4.42}$$

It should be noted that higher order terms are neglected in this solution. The way this result is used is that the plastic zone effectively makes the crack seem slightly longer than it really is. Thus, in fracture calculations an effective crack length is used rather than the actual crack length. This effective length is simply:

$$a_{\text{eff}} = a + \rho \tag{4.43}$$

As long as this zone is small, traditional LEFM assumptions and solutions can be applied using this correction.

Hillerborg approach

Hillerborg *et al.* (1976) used an approach similar to Dugdale and Barenblatt to model the effects of the fracture process zone in concrete. To account for microcracking and bridging, a fictitious crack, or a 'cohesive zone' is proposed. In this model, the tip of the real crack is replaced by an equivalent crack containing closing stresses at the tip. Unlike Dugdale, the stress distribution is not constant, but rather it follows a measurable function of crack opening. That function can nominally be measured in a special uniaxial tension test as shown in Figure 4.18. With three displacement gages, one gage (Gage 2) captures the displacement of the cracked section, while the other two (Gages 1 & 3) capture the displacement of the elastic portions on either side of the cracked section. The total specimen deformation is given by the sum of all gages. However, if all gages are examined individually, it is seen that the gages spanning elastic sections actually close after peak load is reached, while only the gage spanning the crack continues to open. If the elastic portion is subtracted out from Gage 2, a stress-elongation curve for the cracked section is obtained. This can be related to the closing stresses acting at the tip of the fictitious crack.

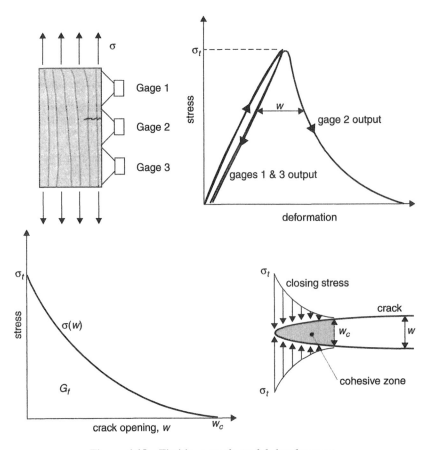

Figure 4.18 Fictitious crack model development.

The fracture energy for such a specimen is determined by taking the area under the stress-crack opening curve:

$$G_F = \int_0^{w_c} \sigma(w)\,dw \qquad (4.44)$$

where G_F is the fracture energy, w_c is the opening of the crack at the start of the stress-free zone, and $\sigma(w)$ is the stress-elongation function. In this model G_F is assumed to be a material parameter, as is the stress-elongation curve. However, it has been experimentally shown that G_F is in fact dependant on specimen size (Bazant and Planas, 1998). Because of the toughening mechanisms in the fracture process zone have a greater influence in smaller specimens, they tend to exhibit higher toughness than larger specimens.

Crack tip opening displacement

The crack tip in a ductile material often has a much different character than that of a brittle material. Wells (1963) observed significant blunting at the tip of existing

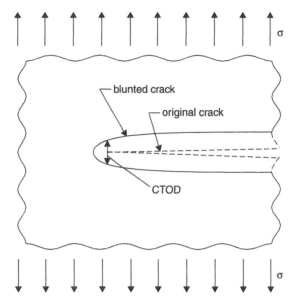

Figure 4.19   Crack tip opening displacement.

cracks as shown in Figure 4.19. Furthermore, he found that tougher materials had a larger degree of blunting prior to rupture. He therefore proposed a Crack Tip Opening Displacement (CTOD) fracture criterion that better reflects the fracture toughness of these ductile materials. He proposed that a critical crack tip opening displacement (CTOD$_C$) could be considered a material parameter and that fracture would initiate when CTOD $=$ CTOD$_C$ in the specimen. Conceptually CTOD can be defined as shown in Figure 4.19, however this is a somewhat ambiguous measurement.

While useful as a laboratory tool, the CTOD criterion has not gained wide acceptance due to the difficulty in both defining a unique CTOD for a given crack tip, and measuring it in real structures. Despite its practical shortcomings, CTOD has relevance in a variety of materials. Mott (1995) observed significant crack tip blunting in cracks extending from pits in individual wood fibers.

4.4.4   The J-integral

Clearly, the above-mentioned effective crack and CTOD approaches have limitations of various forms. Most of these limitations stem from the difficulty in characterising the state of the crack tip in anything but the most homogeneous materials. An alternative approach to the above stress-intensity-based models is the J-integral. J is a path independent counter-clockwise contour integral (illustrated in Figure 4.20) with the following form:

$$J = \int_{\Gamma} \left(w \, dy - \mathbf{T} \frac{\partial \mathbf{u}}{\partial x} \, ds \right) \tag{4.45}$$

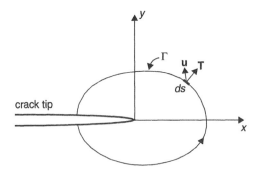

Figure 4.20   J-integral definition of terms.

where Γ is the is the contour of integration, **T** is the stress vector acting perpendicular to the contour, **u** is the displacement vector, and w is the strain energy density, given by:

$$w = \int_0^{\varepsilon} \sigma(\varepsilon)\,d\varepsilon \qquad (4.46)$$

Rice (1968) demonstrated that J is the strain energy release rate for the crack inside the contour, and thus has units of energy per unit area. J is valid for nonlinear elastic materials as well as linear elastic materials. For linear elastic materials $J = G$. J is path independent so Γ can be taken to be any convenient contour around the crack tip. It should be noted that Γ is not closed. It starts on one crack face and ends on the other. $J = 0$ for any closed contour.

J can be a powerful tool for analysing crack problems. For example, many commercial finite element packages contain routines for calculating J for a user-defined contour. The limitation of J is the fact that it is based on an *elastic response* for the material. While inelastic materials clearly violate this assumption, it can be noted that for monotonic loading an inelastic material will typically exhibit a similar stress-strain response as a nonlinear elastic material. The differences occur when the material is unloaded. Then the inelastic material will have permanent deformation, the non-linear elastic material will not. Thus, J is appropriate for monotonic loading conditions where material unloading is not significant. It should be clear that the size of the process zone relative to the contour is important for an inelastic material. Clearly, there is local unloading as the crack advances due to the creation of stress-free surfaces. Since J cannot handle inelastic unloading, the local unloading around the newly formed crack must be small, otherwise the calculation will not be accurate. If the process zone is large compared to the dimensions of the structure, J is not an appropriate fracture parameter.

4.5   Relevance of Material Morphology

It should be clear from topics already discussed that the structure of the material plays a critical role in the fracture criteria set forth. While it should be clear that the microstructure of the material will influence the fracture resistance R, it should also be noted that the structure of the material can influence energy release rate G as well.

As described in Section 4.4.1, there are a number of toughening mechanisms that increase the energy required to propagate a crack. Examples listed included plastic deformation, crack bridging, and microcracking. Clearly, plastic deformation represents a conversion of strain energy to thermal energy, and does not contribute to crack growth. It thus increases the crack resistance, R. Similarly, the formation of microcracks can be thought of as energy dissipated that does not advance the main crack. Hence they too increase resistance, R. Crack bridging, on the other hand can be thought of as affecting both the G and the R side of Equation (4.12). Fibres that bridge the crack and keep it closed simultaneously reduce the energy available to grow the crack (by altering the geometry of a crack that would otherwise open more), and increase the resistance of the material by absorbing additional strain energy. In summary, the microstructure of a material affects fracture toughness by influencing both G and R.

An important feature that arises in many material systems and has a significant influence on fracture toughness is the interface between two similar or dissimilar constituents in a material system. Specifically, the strength of a material interface is known to have a strong influence on fracture toughness. It is now generally recognised that weak interfaces tend to *increase* fracture toughness, while strong interfaces generally decrease fracture toughness.

This perhaps unintuitive result can be attributed to something known as the Cook–Gordon mechanism (Cook and Gordon, 1964). From elastic theory, it can be shown that the stress slightly ahead of a crack tip, in a direction parallel to the crack plane is typically a tensile stress. That is, for a crack oriented along the x-axis, an element located on the x-axis a short distance ahead of the crack tip experiences a horizontal tensile stress σ_x. Because of this, as the crack advances towards an interface, the tensile stresses have the potential to cause a separation of the interface, depending on its strength. As illustrated in Figure 4.21, when a crack approaches a weak interface, the interface will break before the crack tip intersects the interface. Thus, when the crack actually does intersect the interface, the tip becomes blunted, effectively decreasing G (and K), and causing the crack growth to stop. However, as a crack approaches a strong interface, the interface does not separate due to the tensile forces, and the crack continues through the interface as if it was not there. The potential to stop the crack through blunting is eliminated by the strong interface.

Obviously, there is limit to interface weakness. The limiting case of zero interface strength produces a material that could lack both strength and toughness. Clearly, there should be an optimum based on the relative strength and toughness of the different materials that form the interface.

There are numerous examples of this phenomenon in a wide variety of materials, but it is most commonly observed in composite materials. Strong interfaces between fibre and matrix materials typically leads to more brittle behaviour, while weaker interfaces lead to tougher behaviour. In most species of wood, the bonds between wood fibres can generally be considered a weak interface, and as such there is evidence of the Cook-Gordon mechanism at work for loading in tension parallel to grain. It should be noted that weak interfaces have the potential to increase toughness, however this frequently comes at the expense of strength. As observed in many materials, from steel to concrete and ceramics, high strength materials tend to be more brittle (lower toughness) than conventional strength versions of the same group.

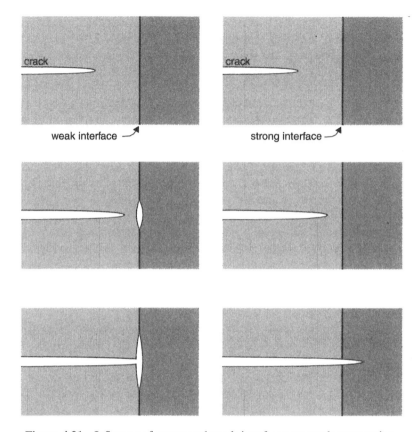

Figure 4.21   Influence of strong and weak interfaces on crack propagation.

4.6   Continuum Damage Mechanics

Damage mechanics is a phenomenological approach that has been developed for materials that do not exhibit plastic deformations, but also cannot be characterised by brittle rupture (Kachanov, 1986). The basic idea is that certain materials exhibit a decrease in stiffness due to the formation of microcracks, or more generically, 'damage'. Damage is quantified by a 'damage variable' that defines the extent to which stiffness has been decreased.

When there is no plastic deformation, the uniaxial stress-strain response of a material subjected to varying degrees of damage can be illustrated as shown in Figure 4.22. In this figure, initial loading of the material can be represented by the curved segment *OAB*. The nonlinearity before peak stress indicates the onset of damage (around point *A*). If the specimen is unloaded at *B*, the stress will follow the straight line *BO*. Upon reloading the stress will again follow line *OB*. No additional damage takes place until B is reached, at which point the stress-strain curve will follow segment *BC*. This process can be repeated with unloading and reloading along segment *CO*. Note that the unloading portions of the curve always return to the origin, meaning no plastic or other inelastic deformation takes place. The change in slope of the loading-unloading diagram represents the accumulating damage. This can be contrasted with

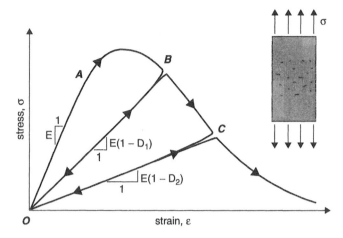

Figure 4.22   Illustration of damage on uniaxial tension specimen.

elastic-plastic deformation where loading and unloading follow lines of parallel slopes (equal to the elastic modulus), where the offset between the parallel lines represents the plastic strain.

Damage can be quantified in terms of the curves illustrated in Figure 4.22. For an undamaged linear elastic material, the stress-strain diagram follows the line:

$$\sigma = E\varepsilon \tag{4.47}$$

A damage variable, D, can be introduced that accounts for the change in stiffness caused by the damage such that:

$$\sigma = (1 - D)E\varepsilon \tag{4.48}$$

where D varies from zero (no damage) to one (complete rupture). In the figure, D_1 denotes the damage that has accumulated up to point B, while D_2 is the damage that has accumulated up to point C.

It should be noted that the description here referred to a uniaxial stress state, where D is a scalar quantity. To generalize damage to a wide range of stress states and damage orientations, D must become a higher order tensor quantity.

4.7   Issues Affecting Applications of Fracture Mechanics to Wood

Fracture mechanics is in many ways an ideal framework in which to study wood failure, and in other ways it is a difficult application. Wood, even at its best, is full of cracks and defects. As fracture mechanics explicitly deals with the relationship between a material's cracks and its strength, the application seems natural. However, much of today's body of fracture mechanics knowledge was developed to solve problems in metals and other relatively 'clean', homogeneous materials. As such, many of the solutions presented in this chapter were derived with a homogeneous, elastic, isotropic solid as a starting point. While this does not preclude application of these solutions to wood, it does suggest that they be applied with a healthy state of caution.

For example, it was previously mentioned that application of a stress intensity factor, K, to a crack problem assumes a nominally sharp crack tip. Casual examination of crack tips in wood under a microscope shows that at a not particularly high magnification, the crack tip gets lost in the complex structure of the material. Indeed, examinations of cracks and crack surfaces show very complex localised failure modes with many different energy dissipation mechanisms at work. As discussed in Chapters 5 and 7, applications of linear elastic fracture mechanics to wood generally fail because of this complex and extensive fracture process zone. Successful applications of fracture mechanics to normal size wood structures must account for these toughening mechanisms. However, many innovative approaches have been, and are continuing to be developed in an effort to move wood mechanics into a more deterministic setting for analysing and predicting structural performance.

4.8   References

Anderson, T. L. (1995) Fracture Mechanics: Fundamentals and Applications, CRC Press, Boca Raton, FL, USA.

Barenblatt, G. I. (1962) 'The mathematical theory of equilibrium of cracks in brittle fracture.' *Advances in Applied Mechanics*, **7**: 55–129.

Bazant, Z. P. (1999) 'Size effect on structural strength: a review.' *Archive of Applied Mechanics*, **69**: 703–725.

Bazant, Z. P. and Planas, J. (1998). Fracture Mechanics and Size Effect in Concrete and Other Quasibrittle Materials, CRC Press, New York, USA.

Broek, D. (1986) Elementary Engineering Fracture Mechanics, Martinus Nijhoff, Dordrecht, The Netherlands.

Cook, J. and Gordon, J. E. (1964) 'A mechanism for the control of crack propagation in all-brittle systems.' *Proceedings of the Royal Society*, **A252**: 508–520.

Dugdale, D. S. (1960) 'Yielding of steel sheets containing slits.' *Journal of the Mechanics and Physics of Solids*, **8**: 100–108.

Daniel, I. M. and Ishai, O. (1994) Engineering Mechanics of Composite Materials, Oxford University Press, New York, USA.

Griffith, A. A. (1921) 'The phenomena of rupture and flow in solids.' *Philosophical Transactions of the Royal Society of London*, **221**: 163–197.

Hillerborg, A., Modeer, M. and Petersson, P. E. (1976) 'Analysis of crack formation and crack growth in concrete by means of fracture mechanics and finite elements.' *Cement and Concrete Research*, **6**: 773–782.

Inglis, C. E. (1913) 'Stresses in a plate due to the presence of cracks and sharp corners,' *Transactions of the Institute of Naval Architects*, **55**: 219–241.

Kachanov, L. M. (1986) Introduction to Continuum Damage Mechanics, Kluwer Academic, Dordrecht, The Netherlands.

Mott, L. (1995) 'Micromechanical properties and fracture mechanisms of single wood pulp fibers,' PhD Dissertation, University of Maine, Orono, Maine, USA.

Rice, J. R. (1968) 'A path independent integral and the approximate analysis of strain concentration by notches and cracks.' *Journal of Applied Mechanics*, **35**: 379–386.

Wells, A. A. (1963) *Application of Fracture Mechanics at and Beyond General Yielding*, British Welding Research Association Report M13.

Williams, J. G. and Ewing, P. D. (1972) 'Fracture under complex stress — the angled crack problem,' *International Journal of Fracture Mechanics*, **8**: 441–446.

Whittaker, D. B. (2002) 'The Application of Acoustic Emission to Measurement of Fracture in Wood,' MS Thesis, University of Maine, Orono, Maine, USA.

USDA. (1999) Wood Handbook: Wood as an engineering material, Forest Products Laboratory, Forest Service, United States Department of Agriculture, US Government Printing Office, Washington, DC, USA.

Appendix: Notation

(primary or recurring items only)

a	= crack length
a_C	= critical crack length
A	= crack area
b	= specimen thickness
C	= specimen compliance
D	= damage variable
CTOD	= crack tip opening displacement
E	= elastic modulus
F	= work of external load
G	= energy release rate
G_C	= critical energy release rate
h	= height of cantilever beam
J	= J-integral
K	= stress intensity factor
K_I	= mode one stress intensity factor
K_C	= critical stress intensity factor
N	= number of loading cycles in fatigue
P	= external load
R	= crack resistance
U	= elastic strain energy
W	= energy of new crack surfaces
β	= dimensionless constant
γ	= specific surface energy
δ	= specimen deformation
ε	= strain
Π	= total energy associated with cracked body
σ	= stress
σ_f	= fracture or failure strength

5

Fracture and Failure Phenomena in Wood

5.1 Fracture and Failure

The structure of wood changes when a sufficient load is applied. Changes that can occur take many forms, and result from a wide range of loading conditions. If the change in structure results in broken bonds and new material surfaces, however large or small, it is referred to here as fracture. Depending on the load condition, fracture can be incremental and gradual, or it can catastrophic and rapid. Failure in this context is simply the point at which wood can no longer carry its load. Failure can result from a number of material mechanisms that are not all necessarily fracture. Likewise, fracture does not always lead to failure. In Chapter 3, failure was discussed in terms of stresses and stress-based strength theory. In Chapter 4, it was shown that strength theory is not appropriate for materials that are sensitive to cracks and the resulting stress concentrations. This chapter focuses on measurements of wood fracture parameters, and the phenomenology of observable fracture processes.

A brief note on semantics is made here. Traditionally, wood engineers make distinctions between clear wood that is nominally free of macroscopic defects (free of visible defects), and massive wood (structural wood) that has an assortment of shakes, checks, knots, and other deviations from the 'clear' condition. In a fracture mechanics context however, there is no need to make distinction, because there is no such thing as defect free wood. One needs only to increase the magnification of observation, and cracks or defects of some kind will eventually come into view. In fracture mechanics theory, a minimum size of crack is not necessary to evaluate an energy release rate or a stress intensity factor. However, implicit in the analysis is that a crack does exist. Fracture mechanics does not deal with initial nucleation of cracks. It deals only with the influence of cracks on macroscopic fracture behaviour. Since cracks or crack-like features exist in wood at a microscopic level from an early age (see Chapter 2), the question of formation of cracks can be dispensed with in subsequent fracture analysis. Thus no distinction is made here between wood with cracks or without cracks.

Fracture and Fatigue in Wood   I. Smith, E. Landis and M. Gong
© 2003 John Wiley & Sons, Ltd   ISBN: 0-471-48708-2 (HB)

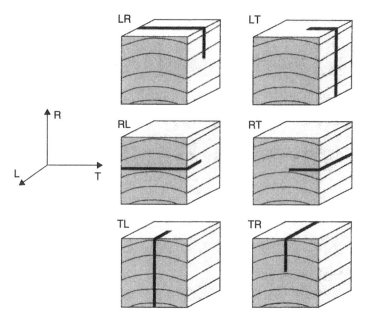

Figure 5.1   Fracture orientations relative to wood growth axis.

Because of the structure that evolves during growth of the tree, the fracture and failure properties are dramatically different depending on how the material is loaded relative to the axis of growth. Six different fracture orientations relative to the growth axis can be defined as illustrated in Figure 5.1. The first letter in the notation of orientation specifies the direction perpendicular to the plane of the crack, and the second letter specifies the direction of crack extension, with L, R and T representing longitudinal, radial and tangential directions respectively. Given that each of these six orientations can, in theory, be subjected to three fracture modes (mode I opening, mode II in-plane shear, mode III out-of-plane shear), there is potentially a large number of fracture cases to consider for each species! Conceptually, the sense in which a crack propagates should also be considered, e.g. away from or towards the pith in the R direction, because as discussed in Chapter 2 properties vary within a growth direction. However, it is rarely necessary in practice to make such fine distinctions. In practice, however, crack propagation along the grain (i.e. RL and TL orientations) is usually the primary focus because of the natural cleavage planes over which cracks can easily propagate.

5.2   LEFM-Based Fracture Toughness Measurements of Wood

The first applications of fracture mechanics principles to wood took place in the 1960s and early 1970s (Atack *et al.*, 1961; Porter, 1964; Schniewind and Pozniak, 1971). Initial work focused on applications of Linear Elastic Fracture Mechanics (LEFM) concepts, even though it appears that some researchers may have recognised early on that there were geometry dependencies that would make LEFM applications problematic. Nevertheless, an effort began to measure and tabulate fracture toughness parameters for a variety of species under a range of fracture orientations. Unfortunately, despite the fact that numerous measurements have been made over the past 30 years, to date there is no standard method for determining fracture toughness in any mode. As with any measured material property, this makes real comparison of data problematic.

Historically, critical stress intensity factor, K_{IC}, has been the preferred fracture toughness measure for wood, probably due to its simplicity in measurement and due to the experience of the contemporary researchers working with metals. Thus, most published fracture toughness data is presented in that form. The last ten years or so has seen a shift to energy-based methods to more easily account for nonlinear fracture behaviour, and to address the difficulty in producing 'pure' fracture modes in wood, the latter being a requirement of stress-intensity factor approaches.

In the sections that follow, efforts to measure and catalogue LEFM-based fracture parameters for different wood species is described. For convenience these have been separated according to mode I, II, III and mixed mode, however it should be clear that fracture in real structures is invariably a complex combination of modes.

5.2.1 Mode I fracture

For most materials, mode I tensile cracking is the critical fracture state, and represents the most dangerous condition. In wood structures this is especially true due to the extreme differences in material properties as a function of grain orientation; strength in the radial and tangential directions being 10–30% of that in the longitudinal direction. Indeed, a wide range of arbitrary stress states can induce a mode I condition along the axis of the grain. Thus mode I fracture parallel to the grain (particularly RL and TL) has received the most attention from engineers and wood scientists.

As mentioned above, the stress intensity factor has been the preferred measure of fracture toughness characterisation. Mode I fracture toughness is determined through tests on specimens for which known solutions of stress intensity factor, K_I, exist. Reviewing the concepts described in Chapter 4, it will be recalled that the stress intensity factor is a measure of the strength of the stress concentration at the tip of a sharp crack in the material. It is a function of the crack length, the applied stress, and the specimen geometry, and can be expressed as:

$$K_I = \beta \sigma \sqrt{a} \tag{5.1}$$

where β is a dimensionless geometry parameter, σ is the far field stress, and a is the crack length. The critical stress intensity factor, K_{IC}, is determined essentially by substituting the fracture stress, σ_f, into Equation (5.1) and solving for stress intensity factor.

$$K_{IC} = \beta \sigma_f \sqrt{a} \tag{5.2}$$

a and β are taken from the particular specimen geometry, the most common of which are presented in Table 4.1. It should be noted that the solutions presented in Table 4.1 were developed for isotropic materials, making their use questionable for

wood. However it has been shown, analytically and experimentally, that the deviations between isotropic and orthotropic solutions are negligible for an RL orientation
(Schachner *et al.*, 2000). Indeed the differences do not become significant until the
LR case is considered, where the isotropic solutions underestimate the stress intensity
factor by 30%.

The advantage of the stress intensity approach is that as long as one uses a specimen
geometry for which there is a known solution for K_I (i.e. β), one needs only to know
the initial notch length, and the load at fracture, in order to calculate K_{IC}. This is
an enormous advantage over, for example, strain energy release rate, G, where in
addition to initial crack length, specimen compliance and change in crack length should
be known.

Figure 5.2 illustrates the most common test specimens for mode I fracture tests on
wood. Of those shown the single edge notched beam seems the most popular (Petterson
and Bodig, 1981; Schniewind and Centeno, 1973) although some researchers have used
variants of the double cantilever beam specimen (Patton-Mallory and Cramer, 1987).

Mode I fracture: tension perpendicular to the grain

As mentioned above, mode I tension perpendicular to the grain (RL, TL, RT and
TR orientations) has received the most attention in fracture mechanics applications
to wood. This is likely due its being the critical failure mode in many structures,

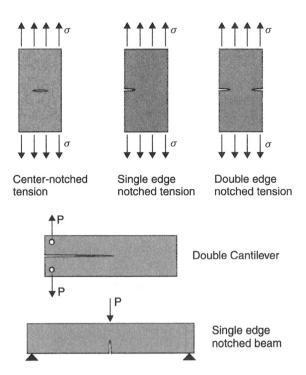

Figure 5.2 Most common specimen geometries for mode I fracture testing of wood.

specifically splitting failure of nailed, bolted and glued connections, and tapered and curved members. On the surface, mode I perpendicular to the grain behaviour appears to be a fairly straightforward problem to fit into an LEFM framework. In terms of the concepts defined in Chapter 4, cracks tend to grow along predefined planes of weakness (lowest crack resistance, R), which coincide with the direction of maximum energy release rate, G. As mentioned above, isotropic solutions for stress intensity factors can be applied with minimal error. Therefore, it is natural that much of the early applications of fracture mechanics to wood took this approach, and applied LEFM principles to this problem.

Table 5.1 presents a list of measured values of K_{IC} for TL and RL orientations in a variety of species in air-dry condition. In the case of Douglas fir, where the results for different test specimen geometry are shown, there is considerable variation in reported values. Higher K_{IC} values are reported when a double cantilever specimen is used as opposed to a single edge notch beam or a double edge notch tension specimen. In early fracture mechanics applications to wood, the reason for this discrepancy was not known (Schniewind and Pozniak, 1971). A basic assumption of LEFM is that fracture toughness (K_{IC}) is a material property that does not depend on specimen geometry. Thus, the measured fracture toughness should be nominally the same in all specimens. While it is possible that some part of the differences are due to normal statistical variations in materials, it seems that the variations are too large to be due to statistical variations alone. What this geometry dependence indicates is that the basic assumption of LEFM applicability is perhaps flawed.

In this context it should be noted that in general, the published K_{IC} values shown in Table 5.1 are based on a single peak load or peak stress value. That is, K_{IC} is determined using equation (5.2), where the failure stress is some predefined value (e.g.

Table 5.1 A sample of measured mode I fracture toughness values for different species[a]

Species	K_{IC} (kNm$^{-3/2}$)		Specimen type
	TL	RL	
Douglas-fir	320	360	
	309[b]	410[b]	single edge notched beam
	260[c]		double edge notched tension
	847[c]		double cantilever beam
Western hemlock	375		
Western white pine	250	260	
Scots pine	440	500	
Southern pine	375		
Ponderosa pine	290		
Red spruce	420		
Northern red oak	410		
Sugar maple	480		
Yellow-poplar	517		

[a]Except where noted, values are taken from the *Wood Handbook* (USDA, 1999).
[b]Schniewind and Centeno (1973).
[c]Schniewind and Pozniak (1971).

peak stress). While convenient from a testing standpoint, fracture measurements based on only a single peak load do not explicitly account for the mobilisation of different toughening mechanisms. Many of the most relevant toughening mechanisms, such as crack bridging and branching are not fully mobilised until the crack shows some extension. Therefore the measured fracture toughness reflects only the initiation of fracture, and neglects the additional energy dissipation that occurs during crack propagation. As detailed in Section 5.3, this assumption underestimates the true toughness of the material, and is the motivation for the development of nonlinear fracture parameters. Since the double cantilever beam specimen produced higher K_{IC} values, it is therefore likely that it promoted conditions for which additional toughening mechanisms could be mobilised.

Another interesting item to consider from Table 5.1 is that for species whose values of K_{IC} are presented for both RL and TL orientations, K_{IC} tends to be higher in the RL compared to the TL orientation. This result is traditionally explained by the fact that cracks in the TL orientation tend to propagate along paths of least resistance caused by rays. Conversely, these same rays can potentially arrest or bridge propagating cracks in the RL orientation, enhancing toughness. Thus the rays can be responsible for both strengthening and weakening.

Relatively little work has been published with regard to fracture in the RT and TR planes, although what has been shown is that fracture toughness is essentially the same in both planes. Schniewind and Centeno (1973) measured K_{IC} to be 355 kNm$^{-3/2}$ (323 psi$\sqrt{in}$) for air dry Douglas fir in both orientations, making them somewhat higher than the TL, but lower that the RL orientation. A likely explanation is that while there is lack of a well defined cleavage plane in the RL and TL orientations, making the crack path somewhat more tortuous, and increasing the energy required to separate the surfaces, there is also a less significant influence of the rays. Regardless of the magnitudes relative to RL and TL orientations, it can be argued that traditional methods of wood construction make this crack orientation less likely to be a troublemaker.

Mode I fracture: tension parallel to the grain

Macroscopically, tension parallel to the grain (LR and LT orientations), where the fracture plane cuts across the wood cells, is not particularly well suited for a traditional Linear Elastic Fracture Mechanics (LEFM) context. The fibrous structure invokes numerous toughening mechanisms that make it an inappropriate application for LEFM: crack bridging, microcracking, and Cook–Gordon crack stopping (see Chapter 4). Thus, it is very difficult to produce a true mode I fracture condition in either the LR or LT direction without a mode II condition arising along the grain. As a result, published values of LR or LT fracture toughness values are somewhat rare. Schniewind and Centeno (1973) published values of K_{IC} for air dry Douglas fir loaded parallel to grain. The median value in the LR direction was 2692 kNm$^{-3/2}$ (2450 psi$\sqrt{in}$), while the median value in the LT direction was 2417 kNm$^{-3/2}$ (2200 psi$\sqrt{in}$). As compared to values listed in Table 5.1, these values are six to seven times greater than their RL, TL, RT and TR counterparts.

Attempts have been made to measure mode I fracture toughness parallel to the grain using a notched beam (Mindess, 1977), and the values obtained for K_{IC} where

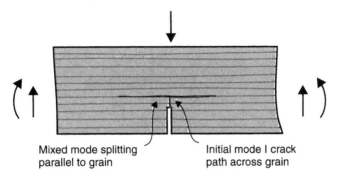

Mixed mode splitting
parallel to grain

Initial mode I crack
path across grain

Figure 5.3 Mixed mode fracture of wood loaded in flexure.

comparable to those obtained using notched tension specimens. However, as illustrated in Figure 5.3, splitting occurs along the grain after an initial crack growth perpendicular to the grain. The Cook–Gordon mechanism is invoked for the initial splitting, but the splitting is propagated by a flexure action not unlike a double cantilever. So while the initial crack growth is clearly mode I, once splitting occurs the exact nature of the fracture mode is open to question.

While the LR and LT orientations have not received the attention of their RL, TL, RT and TR counterparts, their importance should not be diminished. Tension parallel to the grain is frequently the mode of failure in bending elements with knots or other defects. Such failure is typically sudden and catastrophic, and may be considered brittle. Although complete fracture at a structural (massive) scale is not well suited for an LEFM approach, it may be possible to describe the initial crack growth in a linear or nonlinear fracture context if a suitable test method can be devised. In addition, as more work is done at a microscopic level, LR and LT fracture may possibly be characterised in terms of fracture characteristics of wood cells.

Mode I fracture: tension at arbitrary angles relative to the grain

The preceding sections of fracture along the grain and perpendicular to the grain represent limit cases to the issue of arbitrary grain angle. Because of the material anisotropy, it is essentially impossible to get a pure mode I stress intensity at a crack tip in a material with axes inclined relative to the crack plane. In such cases a mixed mode condition arises.

Empirical strength-based relationships have been established to predict failure stress as a function of grain angle, the most well known of which is the Hankinson formula (Hankinson, 1921), which was developed for compressive loads acting at arbitrary angles relative to the grain. It has the following form:

$$N = \frac{PQ}{P \sin^n \theta + Q \cos^n \theta} \tag{5.3}$$

where N is the strength at angle θ measured relative to the longitudinal direction, Q is the strength perpendicular to the grain, P is the strength parallel to the grain, and n is a calibration constant (typically $n = 2$).

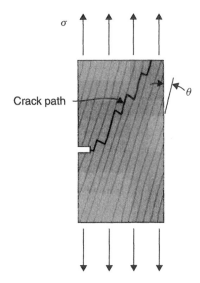

Figure 5.4   Jumping of crack between growth layers (grain boundaries).

In a fracture mechanics context, tensile loading at an arbitrary grain angle is problematic. As has been previously discussed, the issue here is that the directions of least crack resistance, R, and maximum energy release rate, G, will not coincide. The minimum R corresponds to the material (grain) axes, while the maximum G corresponds to direction of principal stress relative to the crack axis. Thus cracks will propagate along the weak axes of the material, but have the potential to periodically 'jump' to the next growth layer (growth ring) in order to relieve the accumulated strain energy, as illustrated in Figure 5.4. This type of nominally straight, but locally irregular, crack propagation has been observed at the microscopic level by Mindess and Bentur (1986). It should be noted that as soon as the crack turns away from the plane perpendicular to the stress field, the fracture becomes mixed mode, which is discussed in more detail below.

5.2.2   Mode II fracture

While mode I tends to receive the most attention, there is no question that mode II fracture plays a significant role in wood structures, particularly for members in flexure and for shear-critical bolted connections. It arguably plays a more critical role than mode I fracture because designers typically work hard (or should work hard!) to avoid situations where tensile stresses perpendicular to grain tend to arise, whereas shear fracture is more likely to be overlooked.

To date, as with mode I, the biggest difficulty in applying fracture mechanics principles to shear failure is the lack of a standardised test to measure fracture toughness. In mode II fracture, the problem stems from the difficulty in producing a pure shear stress state in the material. It is particularly difficult to produce pure mode II stress at the tip of a crack in an anisotropic material. Typically, stresses normal to the crack plane arise that add either a mode I opening component or a closing stress component, either of which

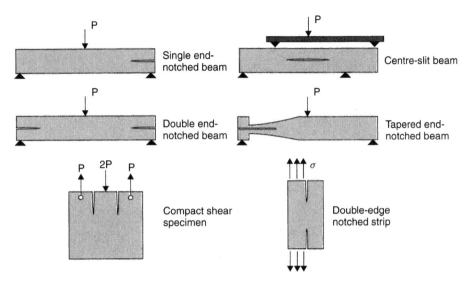

Figure 5.5   Specimen geometries used to measure mode II fracture toughness of wood.

can affect measured results. In fact discussion here is limited to fracture propagating along the grain (RL, TL, RT or TR). Mode II fracture in the LR and LT directions are irrelevant, as there is no practical way to propagate a shear crack across the grain.

Figure 5.5 illustrates the variety of specimens that have been used to characterise mode II fracture toughness. The end-notched beams are more traditional, and have been used to characterise laminated composites for some time. Their drawback has been the frictional and interlocking forces that can arise between the two fracture surfaces during loading due to forces acting normal to the surfaces. These friction forces can be highly variable and significant, causing the measured results to be suspect. Murphy (1988) used the centre-slit beam to reduce the normal forces acting on the crack surfaces. The specimens tend to be somewhat more difficult to fabricate, however. Cramer and Pugel (1987) used a compact shear specimen to eliminate the normal forces altogether. These specimens have the added bonus of offering the potential to be converted to mode I compact tension specimens after failure, assuming of course that shear fracture is confined to one of the two planes.

A tapered end-notched beam has been proposed as a RILEM standard for measuring mode II fracture energy (RILEM Technical Committee 133-TF, 1995). The specimen is designed to produce a stable crack growth under simple displacement control. A Teflon sheet is inserted in the notch over the support point to minimise the sliding friction between notch faces. During the test the specimen is loaded under displacement control to the point where the crack has grown a length of 30–70 mm, where it is unloaded. Fracture energy, G_f, is determined by dividing the work done by the external load by the new crack area formed. The work of load is nominally measured as the difference in areas between the loading and the unloading diagram, as shown in Figure 5.6. The specimen is pried apart in a mode I manner after the test to reveal the fracture surface created under mode II loading. Since mode I and mode II conditions produce significantly different looking fracture surfaces, this last step is done to get a reasonably accurate measure of mode II crack area.

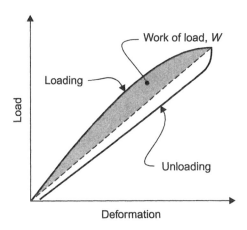

Figure 5.6 Basis of work calculation for mode II fracture energy test.

Table 5.2 A sample of measured mode II fracture toughness values for different species[a]

Species	K_{IIC} (kNm$^{-3/2}$)		Specimen type
	TL	RL	
Douglas-fir		2230	
		1562/1746[b]	center-slit beam
	1370[d]		compact shear specimen
Western hemlock	2240		
	2420/2250[c]		single end-notched beam
Western white pine			
Scots pine	2050		
Southern pine	2070		
	1930[d]		compact shear specimen
Ponderosa pine			
Red spruce	2190	1665	
Poplar		2232[e]	double edge notched tension

[a] Except where noted, values are taken from the *Wood Handbook* (USDA, 1999).
[b] Murphy (1988). Second number indicates results with sharpened notch tip.
[c] Barrett and Foschi (1977). Values shown are for different grades: select structural/No. 2
[d] Cramer and Pugel (1987).
[e] Xu *et al.* (1996).

Table 5.2 presents a sample of published mode II fracture toughness values for a number of different species in an air dry condition. It is more difficult to examine the effects of specimen geometry on measured fracture toughness, but the argument will be made here that, as with mode I testing, there is a geometry dependence. Because of the basic LEFM assumption, all of the researchers specifically cited in Table 5.2 calculated K_{IIC} from a peak load or peak stress measurement. While a valid procedure from an LEFM standpoint, as with mode I, there is no mobilisation of the different toughening mechanisms involved in the fracture process. Since these toughening mechanisms tend to localise near the crack tip at a scale related to inherent microstructural features, size dependence is built into the problem. While the RILEM draft standard is able to capture the influence of these toughening mechanisms, it is likely that it too will produce size dependent results.

Further discussion of size dependence and other non-linear effects are covered in more detail in Section 5.3.

5.2.3 Mixed mode fracture

Mixed mode fracture conditions are particularly interesting because they tend to be the dominant condition for real structures. The general goal of mixed mode fracture is to apply the individually known mode I and mode II critical stress intensity factors K_{IC} and K_{IIC}, to predict the fracture strength under mixed mode conditions. The most common way to produce a mixed mode condition is to put an inclined crack in a uniform tension field such as that shown in Figure 5.7. As discussed in Chapter 4 (Section 4.3.4), in isotropic materials the crack will typically turn so that its plane is perpendicular to the load axis, becoming mode I. By doing this the energy release rate is maximised. However, for a material with predefined planes of weakness such as wood, the crack is more likely to follow the path of minimum crack resistance, R, causing the crack to continue to propagate under mixed mode conditions.

While the individual stress intensity factors can typically be separated and calculated for an arbitrary mixed mode crack tip state, they cannot be separated in terms of the condition for crack growth. That is, crack growth is dependent on not only the individual mode I and mode II fracture toughness values, but also on the interaction between the two. Wu (1967) studied balsa wood under mixed mode conditions using

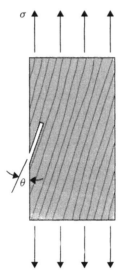

Figure 5.7   Single edge notched specimen for mixed mode fracture testing.

centre-notched plates with the notches cut parallel to the grain axis. Based on experimental results he proposed the following interaction equation as a criterion for mixed mode crack growth:

$$\frac{K_{\mathrm{I}}}{K_{\mathrm{IC}}} + \left(\frac{K_{\mathrm{II}}}{K_{\mathrm{IIC}}}\right)^2 = 1 \tag{5.4}$$

which is frequently cited as a general fracture criterion for anisotropic composite materials. Equation (5.4) is plotted in Figure 5.8. The plot represents a fracture 'envelope' for an arbitrary fracture mode, showing the limit cases for pure mode I and mode II fracture, as well as the mixed mode cases in between. The curve is analogous to the yield surfaces used to predict onset of plastic deformation in metals subjected to arbitrary stress states.

Various investigators have suggested alternative mixed mode fracture criteria of the form:

$$\left(\frac{K_{\mathrm{I}}}{K_{\mathrm{IC}}}\right)^a + \left(\frac{K_{\mathrm{II}}}{K_{\mathrm{IIC}}}\right)^b = 1 \tag{5.5}$$

where a and b are calibration constants. Due, however, to the previously mentioned specimen dependence of fracture toughness values, it is impossible to distinguish superiority of one combination of a and b versus others. In this sense, there can never be a universally valid interaction criteria for mixed mode fracture.

A number of other mixed mode fracture interaction equations have been proposed, including the proposition that mode II has no effect on fracture under mixed mode conditions (shown as the dashed lines in Figure 5.8), meaning no interaction takes place. Mall *et al.* (1983) examined mixed mode fracture of eastern red spruce, and concluded that interaction does occur, and Equation (5.4) is a reasonable model for that interaction.

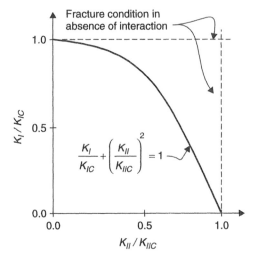

Figure 5.8 Interaction diagram for mixed mode fracture.

5.2.4 Mode III fracture

Mode III fracture (out of plane shear, or tearing) has not traditionally been of great interest to wood engineers, and so there is little published data. Despite this, it is not difficult to imagine mode III applications related to moisture gradient induced fracture or structural connections. It should be noted, however, that mode III is of great practical interest in paper and packaging applications, and as such there is a fair amount of published work in this area (e.g. Swinehart and Broek, 1995).

Ehart *et al.* (1999) developed a novel loading device capable of both mode I and mode III loading. Tschegg *et al* (2001) used the device to examine fracture characteristics of Larch and Beech. Using a fracture energy characterisation (detailed below), they found crack initiation energy in mode III to be over twice as high as mode I in both the RL and TL directions for both species. They attributed this added toughness to a much larger fracture process zone.

5.3 Nonlinear Fracture Characterisation

It should be clear that applications of LEFM theory to fracture of wood have been significant, and that developments have been useful. While there was general agreement among past researchers that reasonable measurements of fracture toughness (as a material parameter) could be made of the materials, and applied to problems of wood failure, it should be apparent that LEFM does not quite account for all the physical phenomena associated with wood fracture. This LEFM theory shortcoming manifests itself in such issues as geometry and rate dependencies of the measured 'material property.' To address the drawbacks, and to better account for the variety of toughening mechanisms involved, researchers started applying nonlinear fracture mechanics methods to fracture processes in wood in the late 1980s.

In Chapter 4, nonlinear fracture mechanics theory is described as a way to account for the effects of a fracture process zone in the vicinity of the crack tip. The necessary condition of LEFM theory is that the influence of this zone is small, and that essentially all available energy goes into the creation of a single new fracture surface. As has been eluded to in the previous sections, and as is further detailed below, numerous material features of wood microstructure cause its behaviour to deviate somewhat from that predicted by LEFM theory. As such, wood may be described as a 'quasi-brittle' material.

A quasi-brittle material is one that is susceptible to some of the same types of catastrophic failures that plague traditional brittle material, however, the behaviour is somewhat less dramatic. As illustrated in Figure 5.9, quasi-brittle materials typically exhibit a nonlinear region prior to the peak load, followed by a strain-softening region after the peak. The term has been applied to a number of non-wood materials that also display these characteristics, namely certain ceramics, concrete, and rock (Shah, 1991).

Fracture behaviour of wood perpendicular to the grain (both RL and TL) can also be characterised as quasi-brittle, and the basis for this behaviour can be attributed to the following microstructural phenomena. As an undamaged[1] specimen is loaded, small

---

[1] 'Undamaged' here refers to a specimen that has not been substantially loaded (beyond routine handling) since it was first cut from the felled log. As has been previously stated, the assumption here is that microscopic cracks and defects are always present.

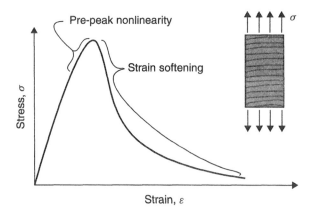

Figure 5.9   Illustration of 'quasi-brittle' behavior of wood in tension perpendicular to grain.

cracks start to grow from microscopic defects in the cell walls and cell boundaries. As these small cracks accumulate, the compliance of the material increases and the load-deformation curve starts to bend over. Prior to peak load there is a localisation process in which the damage that causes failure becomes more confined to a narrow region. By the time peak load is reached, a critical crack accompanied by a fracture process zone has been established, and strain softening can occur. The reason the fracture is not sudden, (the curve falling precipitously), is that toughening mechanisms have been mobilised near the crack tip, causing energy to be dissipated more gradually. An important issue for fracture research is the nature of the process zone, and how it dissipates energy.

Because of the significance of the fracture process zone and its effect on material behaviour, it is necessary to test materials in such a way that the strain softening character may be revealed. That is, it is necessary to test specimens under conditions of stable crack growth. A condition of unstable crack growth will lead to erroneous fracture energy measurements. A popular class of specimen that can produce stable crack growth without too much difficulty is the wedge-splitting configuration illustrated in Figure 5.10a. Variations of this type of specimen have been used since the early days of fracture mechanics research.

In this configuration the nominally rigid wedge is inserted into a pre-cut notch in the specimen. The wedge is pushed into the notch causing splitting along the plane of the notch. The force and the distance the wedge has been pushed into the specimen are recorded. If wedge movement is displacement controlled, then a stable crack growth situation is assured. In terms of the concepts describe in Chapter 4, for this loading configuration a stable equilibrium condition exists when the tip of the wedge is a critical distance from the tip of the crack. When the wedge is pushed further into the specimen, the crack extends in order to maintain that critical distance. Thus stable crack growth is assured.

The area under the wedge force-displacement diagram (Figure 5.10b) is a measure of fracture energy because it is a measure of the work required to grow the crack (Hillerborg, 1991). For generality it is usually expressed in terms of energy per unit

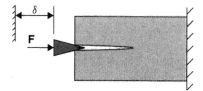

(a) Wedge splitting specimen geometry.

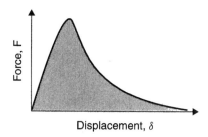

(b) Resulting force - deformation diagram.

Figure 5.10 Illustration of wedge-splitting specimen.

area, and is determined by:

$$G_f = \frac{W}{A} \tag{5.6}$$

where G_f is the specific fracture energy (or just 'fracture energy'), W is the area under the force-displacement curve:

$$W = \int_0^{\delta_{\max}} F(\delta)\, \mathrm{d}\delta \tag{5.7}$$

and A is the newly created crack area. F is the force required to insert the wedge a distance δ. G_f has units of J/m² or N/m. Although similar in character (and units) to critical strain energy release rate, G_C, there is an important difference. Namely G_f is based on the complete separation of surfaces, and it requires the entire load-deformation diagram (including strain softening) to calculate it. G_C on the other hand is based on the energy required to propagate a crack an incremental distance. (As defined in Chapter 4, $G_C = \mathrm{d}U/\mathrm{d}A$ at crack growth.) For a material that has a minimally sized fracture process zone, and LEFM theory applies, then the two quantities are equal. However, for quasi-brittle materials the values can be quite different.

A number of researchers have made comparisons of G_f and G_C, using test results for the same specimen. This is an important comparison because the difference between the two quantities represents the material's deviation from classification as a pure (LEFM-applicable) brittle material. Using measurements made on wedge-splitting specimens of spruce, Stanzl-Tschegg *et al.* (1995) found G_f to be between 10% and 100% greater than G_C in the RL orientation. However they found the differences to be negligible in the TL direction, indicating that LEFM methods are appropriate in that orientation. However, Vasic *et al.* (2002), also using wedge-splitting specimens of spruce found

G_f to be nearly twice as high as G_C in *both* the RL and TL directions, indicating LEFM methods are not applicable in either orientation. In both sets of experiments, G_f was determined by calculating the area under the force-displacement curve for the wedge, while the LEFM solution was determined from a stress intensity factor calculated using finite element methods.

The contradictory results can probably be explained by considering the size of specimen in each group used. The former used specimens that were roughly five times larger than those of the latter, indicating a potential size dependence in the measurements. Indeed, if the results are considered in terms of the size of the fracture process zone, one might hypothesise that the size of the zone is smaller when fracture occurs in the TL plane than when fracture occurs in the RL plane. For the TL specimens of Stanzl-Tschegg *et al.*, it is possible that the specimens were large enough to make the influence of the fracture process zone relatively insignificant, causing the measurements of G_f and G_c to be quite similar. On the other hand, in the smaller specimens of Vasic *et al.*, the process zone was large enough to influence fracture measurements and causing of G_f to be significantly higher than G_c. Further study is necessary to be conclusive on this issue.

This distinction between the LEFM-based parameter G_c and its non-linear counterpart G_f can be examined further in terms of crack initiation energy and crack propagation energy. That is, the energy required to initiate a crack, and to grow the crack once the toughening mechanisms have been mobilised. Tschegg *et al.* (2001) examined this distinction using specially prepared thin, notched specimens of beech and larch loaded in direct tension (both RL and TL orientations). The specimens were loaded in an experimental configuration that allowed stable crack growth to complete separation of specimen pieces. Load and Crack Mouth Opening Displacement (CMOD) were measured for each specimen. In the analysis of their experimental data, they plotted load versus CMOD and calculated fracture energy, G_f, by taking the area under the load-CMOD curve and dividing by the area of the fracture surface. In addition, they separated G_f into initiation and propagation components as illustrated in Figure 5.11. Crack initiation energy, G_{init}, was defined as the area under the load-CMOD curve from the origin to the peak, and back down a line parallel to the initial slope of the curve. The propagation energy, G_{prop}, is simply the difference between the total fracture energy, G_f, and the initiation energy G_{init}.

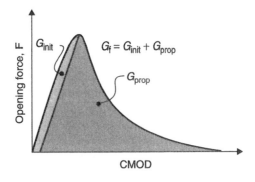

Figure 5.11   Representation of crack initiation versus crack propagation energies.

This crack initiation energy may be interpreted as the energy required to develop a fracture process zone. That is, the energy dissipated through the formation of microcracks that ultimately localise into the main crack. This interpretation seems reasonable in a damage mechanics context. It is logical to assume that the pre-peak nonlinearity observed in the load-CMOD curve is the direct result of microcrack formation. It should be noted, however, that from *in situ* Scanning Electron Microscopy studies of fracture processes Vasic (2000) found very little evidence of activated microcracks in the vicinity of the crack tip. In fact, she concluded that the majority of the energy dissipated during fracture processes was due to bridging behind the crack tip, rather than microcracking ahead of the crack tip. Thus, the exact nature of energy dissipation in the fracture process zone in wood is still being debated.

While there are no parallels to initiation and propagation energy in LEFM, it could be argued that measurements of wood fracture toughness made using LEFM theory are really measurements of crack initiation fracture toughness, and that G_c is nominally equivalent to G_{init}. Published values of G_f, are not particularly abundant at this time, however some sample values for several species are shown in Table 5.3. As with all published values of wood fracture parameters of any sort it must be remembered that no testing standards exist. While most researchers make extensive efforts to conduct their tests in a reasonable manner, variations due to moisture, load rate, specimen size, and other factors will significantly affect the outcome. Comparisons between tests at different laboratories must therefore be viewed with caution.

On the issue of the influence of specimen size, at this point in time there has not been a rigorous study of size effects on fracture toughness or fracture energy measurements of wood. Stanzl-Tschegg *et al.* (1995) considered the effect of both specimen thickness and ligament length[2] on their fracture energy measurement of spruce. For RL specimens ranging in thickness between 10 and 80 mm and ligament length between 30 and 100 mm, they found that specimen thickness had a significant effect on fracture energy, increasing from 130 to nearly 240 J/m². This variation was in specimens ranging from 10–40 mm thick. Above 40 mm the effect of thickness was negligible. Ligament length also had a measurable influence on fracture energy, steadily increasing from 150–250 J/m² over a change in length from 30–100 mm. Their conclusion was that fracture specimens should have thickness greater than 30 mm and ligament length greater than 70 mm for size-independent measurement of fracture energy. It must be

Table 5.3 Sample of measured fracture energy, G_f, for different species[a]

Species	Fracture Energy G_f (J/m²)
Spruce	260
Pine	346
Alder	177
Oak	234
Ash	365

[a]Reiterer *et al.* (2000).

[2] Ligament length in this context is the length of the intact portion of the specimen measured in the plane of the crack.

noted that true measurement of size effect must be done on geometrically similar specimens (Bazant, 1999), so the conclusions can be questioned.

The issue of size effect is important, because as discussed in Chapter 4, materials that exhibit toughening mechanisms tend to have size effects with respect to toughness. The fracture process zone in which the toughening mechanisms act tends to remain nominally constant in size for a given material structure. As specimens get larger and larger, the effects of the toughening mechanisms diminish, leading to more brittle behaviour. Therefore, tabulated values of fracture toughness must be taken with some caution, as they are likely to be size dependent. By quantifying the effects of the toughening mechanisms (in this case dominated by crack bridging stresses), there is hope for moving toward a better representation of true fracture toughness.

5.3.1 Other fracture characterisations

A few researchers have examined the use of the J-integral to characterise fracture of wood. The J-integral is introduced in Chapter 4 as a way to express the energy release rate of a cracked body that has a nonlinear elastic response. Thus, it has been popular in characterising metals where there is significant plastic yielding at the tip of the crack. The pre-peak nonlinearity frequently observed in wood loaded perpendicular to grain, makes the J-integral an attractive method for characterisation of fracture.

Riipola and Fonselius (1992) examined ways to apply an American Society for Testing and Materials developed method for elastic-plastic fracture in metals to the problem of determining J_C in spruce and pine. They found a rate dependence on the measured value. Specifically, slower load rates yielded lower values of J_C. Yeh and Schniewind (1992) used the J-integral to characterise fracture energy in Douglas fir and Pacific mandrone loaded in mode I. They looked at the influence of temperature and moisture content on measured values of J_C, G_C and K_{IC}, and they examined the relationship between the quantities. They concluded J_C was the preferred characterisation, particularly for high moisture contents where ductility is more significant.

As with the other fracture characterisations, these measurements of J_C are likely to be size dependent.

5.4 Material Issues Affecting Fracture Properties

As detailed in Chapters 2 and 3, wood properties are influenced by material features over an extremely large range of length scales: from the macroscopic growth features, all the way down to the molecular level. In a fracture mechanics context the story is no different. Heterogeneity acts over this very large range of length scales, causing the wide distribution of strength properties that have come to be expected from wood. At massive scale the heterogeneity affecting fracture is obvious: shakes and checks are natural starting points for crack growth. Knots, while not crack-like in shape or character, represent locations of stress concentration at which cracks can extend.

5.4.1 Microstructural issues

At a micro-scale the story becomes more interesting, for even in so-called 'clear wood' there is significant heterogeneity. This heterogeneity manifests itself through cell overlap, pits and pores, earlywood-latewood transitions, and ray crossings. All of these features represent locations of stress concentration that influence the measured fracture properties. In addition, cell geometry and species dependent arrangement of cells, are responsible for the mode of microfracture, and therefore have significant affect on overall fracture toughness.

Microscopically, the strength of individual softwood cells can be up to three times that of corresponding early wood cells (Jayne, 1960), with mean strengths of about 500 MPa. At this scale, three types of failure can be distinguished: intercell, intrawall, and transwall, as illustrated in Figure 5.12. Intercell failure (IC) occurs at the middle lamella and represents the separation of cells. Intrawall failure (IW) refers to failure within the secondary wall, most often at the secondary wall S2/S1 interface. When the fracture path cuts across the wall, the failure is described as transwall (TW). In hardwoods, failure results in an extremely complex fracture surface with transwall failure that follows the S2 microfibril angle in cells having a thick S2 layer. Within some cells, there can be intrawall failure in the S1 layer, which causes the fibre core to pull out. The failure pattern in softwoods is similar to that for hardwoods. Thick-walled latewood tracheids fail by transwall fracture that follows the S2 microfibril orientation. In earlywood tracheids, abrupt transwall fracture is typical. Thick walled

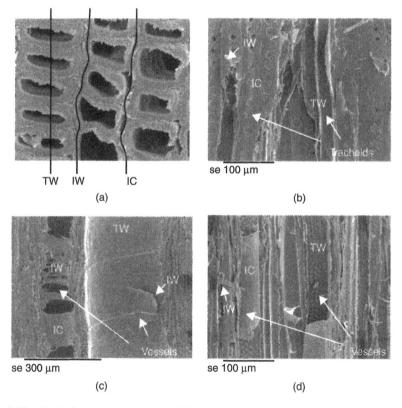

Figure 5.12   Basic fracture paths in wood: intercell (IC), transwall (TW) and intrawall (IW); (a) illustrated fracture paths; (b) fracture in spruce; (c) fracture in oak; and (d) fracture in maple.

tracheids fracture in an abrupt transwall pattern, or S2 layers are pulled out of the S1 layers and are then unwound.

From a fracture toughness standpoint, observations of fracture at the micro scale indicate that wall thickness has a stronger influence on the fracture toughness of individual cells than cell size. Therefore fracture in low density wood tends to be transwall, while for higher density wood both transwall and intercellular fracture is observed. The result of this is that there are frequently different fracture modes in earlywood and latewood cells. Specifically, intercellular fracture is observed in thick-walled latewood cells, while transwall fracture is observed in thin-walled earlywood cells. Clearly the proportion of earlywood to latewood will affect bulk fracture toughness measurements.

It must be emphasised that at these small scales, it is irrelevant to talk in terms of a constant fracture toughness, K_{IC}, or a crack resistance R (or G_C). These are quantities that apply to the macroscopic, or bulk material scale, and are not appropriate for the heterogeneous structure at small scales.

5.4.2   Other issues affecting fracture toughness

As one might expect (as with nearly all useful properties of wood), there are many condition factors that affect the measured fracture toughness of different wood species. A comprehensive review of such factors is probably futile, but there are several that stand out as being particularly significant: moisture and drying history, and temperature. It is noted that these factors go beyond those already discussed, such as species, orientation (e.g. RL, TL) and grain angle.

Moisture content and drying history has a very significant effect on fracture toughness, although the exact nature of that effect appears to be greatly species dependent, and dependent on the extractive content (Petterson and Bodig, 1983). Smith and Chui (1994) evaluated the effect of moisture content on measured fracture energy, G_f, for red pine, and found it reached a maximum at a moisture content of 18%. Indeed, the differences were significant ranging from 455 J/m^2 in its green state, to 565 J/m^2 at 18%, and then down to 345 J/m^2 at 7%. In most materials, ranging from metals to composites, there is a trade-off between strength and toughness. That is, high strength materials tend to be more brittle. In this particular case, the wood follows this trend, but only at moisture contents below 18%.

The effect of drying history manifests itself primarily through shrinkage cracking, producing numerous stress concentrations for crack growth. While shakes and checks are the most visible damage, it is likely that damage at a microscopic level affects measured fracture properties as well.

Temperature also affects measured fracture properties by altering the microfracture mode in the material. For example, Atack *et al.* (1961) found that for black spruce at room temperatures, both transwall and intrawall microfracture modes occur, and resulting fracture surfaces are quite rough. However, at 100°C the microfracture was predominately intercellular, and as a result the fracture surfaces were quite smooth. This change was attributed to plasticisation of the lignin in the middle lamella. Reiterer (2001) evaluated both K_{IC} and G_f for spruce and beech at temperatures ranging from 20° to 80°C. Both values decreased over that temperature range. The biggest drop in K_{IC} occurring between 20° and 40°, while the biggest drop in G_f occurring between 60° and 80°.

5.5   Issues for Material Modelling

All the issues of the preceding sections present interesting challenges for the development of models that can ultimately allow knowledge of material properties to be applied in design of more reliable structures. While this is the topic of Chapter 7, it is worth highlighting a few of the modelling issues in the context of the material properties discussed in this chapter. It should be noted here that implicit in this discussion is that the use of LEFM methods is suspect. If LEFM is a valid model for wood fracture, then implementation is fairly straightforward. Once the relevant critical stress intensity factors are known for a species, one can evaluate fracture strength of a structure by applying the appropriate stress intensity factors to the structure. While this is not always trivial, it does not require additional model development.

Based on the current state of wood fracture knowledge, it is proposed that basic fracture models for wood should be able to accommodate the following materials issues:

- *Multi-scale effects.* The extremely wide range of material length scales affecting fracture properties poses a difficult modelling problem. The traditional computational modelling is problematic because the detail required to capture small scale phenomena quickly makes the modelling of a large scale structure intractable. New paradigms are probably required.

- *Geometry dependencies.* The relationship between the microstructure of the material and the fracture process zone leads to an inherent size effect in fracture energy measurements. This issue affects both measurements of material properties and prediction of structural capacity.

- *Species issues.* Clearly, the differences fracture processes among different species pose a great problem to model generality. Not only is the variation between species considerable, but the variation within species is significant. Moisture state, temperature, and history will all impact the ability of a model to make reasonable predictions.

While great progress has been made in addressing these issues, it is clear that the development of material models to predict fracture properties is a wide open field.

5.6   References

Atack, D., May, W.D., Morris, E.L. and Sproule, R.N. (1961) 'The energy of tensile and cleavage fracture of black spruce' *Tappi*, **44** (8): 555–567.

Barrett, J.D. and Foschi, R.O. (1977) 'Mode II stress-intensity factors for cracked wood beams', *Engineering Fracture Mechanics*, **9**: 371–378.

Bazant, Z.P. (1999) 'Size effect on structural strength: a review', *Archive of Applied Mechanics*, **69**: 703–725.

Cramer, S.M. and Pugel, A.D. (1987) 'Compact shear specimen for wood mode II fracture investigations', *International Journal of Fracture*, **35**: 163–174.

Ehart, J.A., Stanzl-Tschegg, S.E. and Tschegg, E.K. (1999) 'Mode III fracture energy of wood composites in comparison to solid wood', *Wood Science and Technology*, **33**: 391–405.

Hankinson, R.L. (1921) 'Investigation of crushing strength of spruce at varying angles of grain', *Air Service Information Circular*, **3**(259).

Hillerborg, A. (1991) 'Application of the fictitious crack model to different types of materials', *International Journal of Fracture*, **51**: 95–102.

Jayne, B.A. (1960) 'Some mechanical properties of wood fibers in tension" *Forest Products Journal*, June, 316–322.

Mall, S., Murphy, J.F. and Shottafer, J.E. (1983) 'Criterion for mixed mode fracture in wood', *Journal of Engineering Mechanics*, **109**(3): 680–690.

Mindess, S. (1977) 'The fracture of wood in tension parallel to the grain', *Canadian Journal of Civil Engineering*, **4**(4): 412–416.

Mindess, S. and Bentur, A. (1986) 'Crack propagation in notched wood specimens with different grain orientations', *Wood Science and Technology*, **20**: 145–155.

Murphy, J.F. (1988) 'Mode II wood test specimen: beam with center slit', *Journal of Testing and Evaluation*, **16**(4): 364–368.

Patton-Mallory, M. and Cramer, S.M. (1987) 'Fracture mechanics: a tool for predicting wood component strength', *Forest Products Journal*, **37**(7/8): 39–47.

Petterson, R.W. and Bodig, J. (1983) 'Prediction of fracture toughness of conifers', *Wood and Fiber Science*, **15**(4): 302–316.

Porter, A.W. (1964) 'On the mechanics of fracture in wood', *Forest Products Journal*, **14**(8): 325–331.

Reiterer, A. (2001) 'The influence of temperature on the mode I fracture behavior of wood', *Journal of Materials Science Letters*, **20**: 1905–1907.

Reiterer, A., Stanzl-Tschegg, S.E. and Tschegg, E.K. (2000) 'Mode I fracture and acoustic emission of softwood and hardwood', *Wood Science and Technology*, **34**(5): 417–430.

RILEM Committee 133-TF Fracture of Timber (1995) 'Draft recommendation for determination of fracture energy in forward shear along the grain in wood', *Materials and Structures*, **28**: 482–487.

Riipola, K. and Fonselius, M. (1992) 'Determination of critical J-integral for wood', *Journal of Structural Engineering*, **118**(7): 1741–1750.

Schniewind, A.P. and Centeno, J.C. (1973) 'Fracture toughness and duration of load factor'. I. Six prinicipal systems of crack propagation and the duration factor for cracks propagating parallel to grain, *Wood Fiber*, **5**(2): 152–159.

Schniewind, A.P. and Pozniak, R.A. (1971) 'On the fracture toughness of Douglas fir wood', *Engineering Fracture Mechanics*, **2**(3): 223–233.

Schachner, H., Reiterer, A. and Stanzl-Tschegg, S.E. (2000) 'Orthotropic fracture toughness of wood', *Journal of Materials Science Letters*, **19**(20): 1783–1785.

Shah, S.P. (1991) Toughening Mechanisms in Quasi-brittle Materials, Kluwer Academic, Dordrecht.

Smith, I. and Chui, Y.H. (1994) 'Factors affecting mode I fracture energy of plantation-grown red pine', *Wood Science and Technology*, **28**: 147–157.

Stanzl-Tschegg, S.E., Tan, D.M. and Tschegg, E.K. (1995) 'New splitting method for wood fracture characterization', *Wood Science and Technology*, **29**: 31–50.

Swinehart, D. and Broek, D. (1995) 'Tenacity and fracture toughness of paper and board', *Journal of Pulp and Paper Science*, **21**(11): J389–J397.

Tschegg, E.K., Frühmann, K. and Stanzl-Tschegg, S.E. (2001) 'Damage and fracture mechanisms during mode I and III loading of wood', *Holzforschung*, **55**(5): 525–533.

USDA (1999) Wood Handbook: Wood as an engineering material, Forest Products Laboratory, Forest Service, United States Department of Agriculture, US Government Printing Office, Washington, DC, USA.

Vasic, S. (2000) 'Applications of fracture mechanics to wood', PhD Thesis, University of New Brunswick, Fredericton, NB, Canada.

Vasic, S., Smith, I. and Landis, E. (2002) 'Fracture zone characterization — micro-mechanical study', *Wood and Fiber Science*, **34**(1): 42–56.

Wu, E.M. (1967) 'Application of fracture mechanics to anisotropic plates', *Journal of Applied Mechanics*, **34**(4): 967–974.

Xu, S., Reinhardt, H.W. and Gappoev, M. (1996) 'Mode II fracture testing method for highly orthotropic materials like wood', *International Journal of Fracture*, **75**: 185–214.

Yeh, B. and Schniewind, A.P. (1992) 'Elasto-plastic fracture mechanics of wood using the J-integral method', *Wood and Fiber Science*, **24**(3): 364–376.

Appendix: Notation

(primary or recurring items only)

a = crack length
A = crack area
CMOD = crack mouth opening displacement
G = energy release rate
G_C = critical energy release rate
G_f = specific fracture energy (or just "fracture energy")
G_{init} = crack initiation energy
G_{prop} = crack propagation energy
J = J-integral
K = stress intensity factor
K_i = mode i stress intensity factor
K_{iC} = critical stress intensity factor for mode i
L = longitudinal direction
LEFM = Linear Elastic Fracture Mechanics
R = crack resistance
R = radial direction
T = tangential direction
U = elastic strain energy
W = energy of new crack surfaces
β = dimensionless geometry parameter
σ = far field stress
σ_f = fracture stress

6

Fatigue in Wood

6.1 The Phenomenon

Fatigue is the process of progressive localised permanent structural change occurring in materials subjected to conditions that produce sustained or fluctuating stress and strains at some point or points, and that may culminate in cracks or complete fracture after time or sufficient number of fluctuations.

Fatigue failure of a material results from sustained or cyclic application of stress less than is required to cause in-elastic behaviour or fracture under monotonic loading conditions. Damage initiates as micro cracks that subsequently aggregate, leading eventually to macro cracking and failure. The definition of fatigue given above is similar to that employed by the American Society for Testing and Materials (ASTM, 1993), but incorporates a time argument because of wood's rheological nature. Any material exhibiting flow behaviour must exhibit fatigue that depends on the number of stress cycles, and the rate of stressing and/or time under stress. Unfortunately, this has not been widely recognised in the past.

The lay user of wood knows that heavily loaded shelves sag over time and that barns and roofs can collapse after several days of heavy snow loading. Modern engineers accept as fact that sustained stress has a damaging (static fatigue) effect on wood. Over 260 years ago the French naval architect Georges Louis Le Clerc, Compt de Buffon stated that the ability of wood to carry load indefinitely depends on the stress level (σ), and recommended that σ for oak beams not exceed 0.5 (Buffon, 1740). Generations of boat and ship builders have known that some wood species have better fatigue performance than other species. However, the famous World War I aircraft engineer Dr. Fokker is reported to have believed wood immune to fatiguing effects of cyclic stress (Lewis, 1960; Kyanka, 1980). Most modern engineers follow his lead giving no or scant explicit consideration to the possibility of cyclic fatigue' damage. Such thinking supposes that 'allowable design stresses' are below a threshold where repeated stresses would cause damage (ASCE Committee on Wood, 1975).

Catalysts for recent study of fatigue in wood members and connections include need to design wind turbines from wood or wood-plastic composite materials (Tsai

Fracture and Fatigue in Wood I. Smith, E. Landis and M. Gong
© 2003 John Wiley & Sons, Ltd   ISBN: 0-471-48708-2 (HB)

and Ansell, 1990), and anti-seismic design of wood structures (Smith *et al.*, 1998). These two examples are at extreme ends of the fatigue spectrum. Windmill blades are deliberately oriented to maximise wind force in order that as much energy as possible is extracted from the wind. They must resist millions of cycles of fairly low level stress. This is termed High Cycle Fatigue (HCF) behaviour. Seismic damage is by contrast due to rare events of very short duration, with the objective being to resist relatively few load cycles that develop very high levels of stress. This is termed Low Cycle Fatigue (LCF) behaviour. Although there is no precise boundary between HCF and LCF, it has been suggested that HCF is when failure is caused by more than 10 000 stress cycles (Puskar and Golovin, 1985; Ansell, 1995).

The remainder of this chapter explores fatigue concepts and experimental knowledge of fatigue in wood. Modelling and prediction of fatigue damage is dealt with in Chapters 8 and 9.

6.2   State of Experimentally Based Knowledge

6.2.1   Overview

Traditionally attention has mainly been on fatigue damage in wood caused by sustained stress, with relatively little attention given to effects of cyclic stress. This is because, as already mentioned, the duration of load effect on strength has been perceived as the dominant concern for engineers (Section 6.1). *Ad hoc* test observations reveal that time to failure T under sustained stress is related to the applied stress level σ, with T being reduced approximately exponentially due to any proportional increase in σ. The so-called 'Madison curve' (Wood, 1951) provides empirical underpinning to design level adjustments accounting for the phenomenon (Figure 6.1). Structural wood design codes throughout the world employ the Madison curve, or a more recent substitute empirical relationship, to link permitted levels of stress with cumulative duration of transient peak design forces. Codified design practices includes no explicit recognition of the possibility of cyclic fatigue. This reflects the argument that any cyclic load induced fatigue is allowed for implicitly because 'cumulative lifetime duration of peak force' is adopted in conjunction with Madison type static fatigue relationships. Thus, the engineering level premise is that time under load is always the controlling influence on accumulation of fatigue damage. As discussed below, this is not always a reliable presumption.

Over recent years there has been *ad hoc* investigation of cyclic load induced fatigue in wood. Investigation has focussed on relatively rapidly applied load because of the necessity to obtain information in a reasonable span of time. However, this compromises ability to extrapolate findings because with wood, and other rheological materials, it is not possible to accelerate cyclic load tests by simply increasing the loading frequency (Nielsen, 1996). Length of experiment constraints are especially limiting in the context of proprietary 'wood with glue' composites the makeup of which is often modified quite frequently. As wood absorbs water in both the liquid and vapour forms, the amount of moisture in it fluctuates with change in the relative humidity and temperature of the surrounding medium (usually air). Fatigue properties are affected by the amount of water present in wood. However, the extent of moisture sensitivity following initial drying depends upon member dimensions because moisture movement is a time

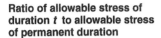

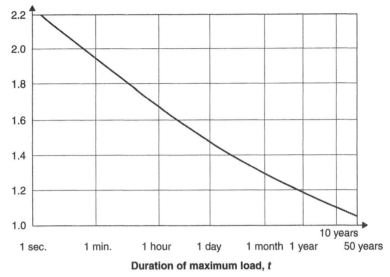

Figure 6.1 'Madison curve': relationship of working stress to duration of loading.

dependent transportation process. Judged at the system level, large members tend to be more prone than small members to damage during initial drying, but less prone to stress induced fatigue damage during moisture cycling because it usually only effects outer layers. Interpretation of data needs to recognise the rheological and hygroscopic nature of the material. Although it is possible to accelerate fatigue tests by studying behaviour at elevated temperatures (Mohammad and Smith, 1994, 1996), this is normally not done with wood because the physics of the process are not well understood.

Even though information on fatigue mechanisms and processes is fragmented and often sparse, certain aspects of behaviour of wood-based materials and connections in such materials are quite widely (if not unanimously) accepted:

- For repetitive cyclic stress a fatigue limit is thought to exist for wood (Kyanka, 1980; Ansell, 1995) and connections (Hayashi *et al.*, 1980).[1] There is dissenting opinion on this (Rosenthal, 1964), but on balance it seems acceptable as a working hypothesis. 'Damaged visco-elastic material theory' predicts that materials such as wood will fail irrespective of a fatigue limit at large times to failure (Nielsen, 1996).

- The number of repetitive load cycles to failure wood components is smaller under reversed (through-zero) cyclic stress than under non-reversed (one-side-of-zero) cyclic stress having the same peak stress (Tsai and Ansell, 1990). The most damaging situation is fully reversed loading, i.e. stress cycles for which the mean stress is zero. Nielsen (1996) proposes the existence of a critical ratio of negative to positive stress under reversed loading, R_{crit}, above which the effects of stress-reversal

---

[1] A fatigue limit is a level of stress at or below which repetitive stress may be applied indefinitely without failure.

are negligible. His order of magnitude estimate is R_{crit} equals -0.5. However, this value appears inconsistent with the findings of Tsai and Ansell (1990), who studied flexural behaviour of sliced African mahogany Laminated Veneer Lumber (LVL). Their work implies that R_{crit} is close to zero. Possibly the value depends upon the strength level of a material, where strength level *FL* is the ratio of the actual strength to the theoretical strength of undamaged wood substance (lumber and glulam have lower *FL* than clear wood).

- The number of repetitive load cycles to failure (fatigue life), *N*, decreases as the moisture content is increased between 5% and the fibre saturation point (Tsai and Ansell, 1990). Moisture appears to have a simple shift effect on the fatigue life as represented on stress level vs. fatigue life ($S - N$ or $\sigma - N$) diagrams (Figure 6.2) (Tsai and Ansell, 1990; Nielsen, 1996).

Figure 6.3 illustrates how stress versus number of load cycles to failure ($\sigma - N$) diagrams for wood compare with those for other common construction materials,

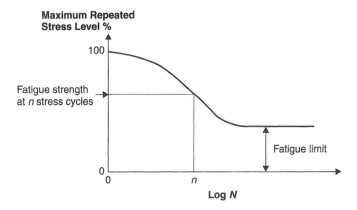

Figure 6.2   Stress level versus number of cycles to failure ($S - N$) curve.

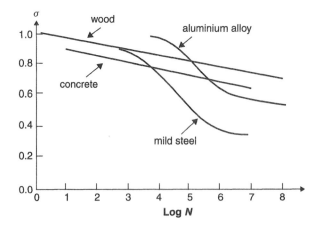

Figure 6.3   Stress level versus fatigue life ($\sigma - N$) diagrams for mild steel, aluminium alloy, concrete and wood (schematic).

Table 6.1 Selected fatigue life data (based on FPL 1989)

Property	Stress ratio, R-ratio	Loading frequency (Hz.)	Peak stress level, σ	Approximate fatigue life, N (cycles $\times 10^6$)
Bending:				
cantilever	0.45	30	0.45	30
cantilever	0.0	30	0.40	30
cantilever	−1.0	30	0.30	30
– centre-point	−1.0	40	0.30	4
rotational	−1.0	—	0.28	30
third-point	0.1	8.33	0.60	2
Tension parallel to grain:				
	0.1	15	0.50	30
	0.0	40	0.60	3.5
Compression parallel to grain:				
	0.1	40	0.75	3.5

based on Hansen (1991). The *Handbook of Wood and Wood-Based Materials for Engineers, Architects and Builders* (FPL, 1989) summarises results of HCF tests on wood. Table 6.1 gives illustrative fatigue life data for dry clear wood tested in 'normal' room conditions.[2] This data suggests N is about 10^7 cycles at $\sigma = 0.40$, which clearly does not agree with the values implied by Figure 6.3. Such a discrepancy is not unusual, and serves to illustrate that characterising fatigue performance of wood is no simple matter.

6.2.2 Effect of loading variables

Although all loading can be thought of as cyclic, here cyclic loads (stress) are taken to mean those that have loading frequencies sensibly measured in Hz. Damage accumulation in wood under cyclic stress depends upon the work done per cycle and the waveform employed. Shape of a waveform governs important factors such as stressing rate and period of residence at the peak stress. Figure 6.4 illustrates some waveforms having the same peak stress and loading frequency. Segment AB reflects the stressing rate, and segment BC the residence period. Points A, B, C and D coincide with peak rates of change in stress for square and triangular waveforms. In triangular and sinusoidal waveforms, points B and C overlap and the residence period is zero. Approaching an ideal square waveform usually is not possible and in practice waveforms are only approximately square cornered (bottom-right diagram).

Experiments reveal that for clear wood square waveforms are the most damaging (Okuyama *et al.*, 1984; Gong and Smith, 1999). This is because they have the highest stressing rate, the highest change in the stressing rate and the longest residence period (Table 6.2). Symmetric triangular waveforms are less damaging than sinusoidal waveforms, even though the rate of change in stressing rate is not zero at point B (Table 6.2). This is because peak stressing rate, rather than the rate of change in stressing rate, is

---

[2] Stress Ratio, also known as the R-ratio, is the ratio of the minimum stress to maximum stress. For example, R-ratio = −1.0 represents fully reversed cyclic loading, and R-ratio = 1.0 sustained load.

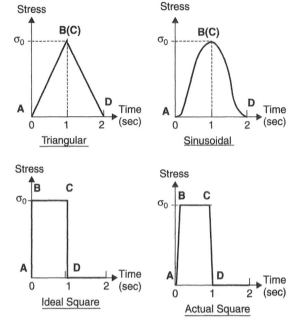

Figure 6.4   Waveforms with loading frequency 0.5 Hz.

Table 6.2   Characteristics of waveforms in Figure 6.4

Waveform	Peak stressing rate (MPa/sec)	Rate of change in stressing at Points B & C	Residence period (sec)
Triangular	$\pm\sigma_0$	$\pm\infty$	0
Sinusoidal	$\pm1.57\sigma_0$	0	0
Square (ideal)	$\pm\infty$	$\pm\infty$	1
Square (actual)	$\pm20\sigma_0$	$\pm\infty$	~1

the predominant factor in damage accumulation under cyclic load. Alternative explanations have been suspected, such as that the energy contents of waveforms relate to the extent of damage, but this is not the case. Although the average stressing rate is equal to σ_0 for both sinusoidal and triangular waveforms, it is not constant for a sinusoidal waveform with the maximum rate $1.57\sigma_0(=\pi\sigma_0/2)$. This is why sinusoidal waveforms are more damaging than symmetric triangular ones of the same amplitude and loading frequency. It should be noted however that the rate of displacement and by implication the rate of loading correlates negatively with the rate of damage accumulation in bolted wood connections (Daneff, 1997). Conclusions related to behaviour of wood alone cannot necessarily be extrapolated to the effects that different waveforms have on fatigue behaviour of wood connections or structural systems.

Often cyclic loads are not repetitive, i.e. cyclic stress or deformation does not have constant amplitude or frequency. It is necessary therefore to know how irregularity in loading sequences affects accumulation of damage. Some insights exist, even though this issue has not been fully explored and knowledge is rudimentary. Gong (2000) used two simple stepped-load sequences (Figure 6.5) to apply compression parallel to grain to clear spruce specimens.[3] Each sequence employed four 'blocks' of 50 stress cycles. The first 45 cycles within a block were at the target σ, and the remaining five were a phased transition to the next σ. Accumulated deformation AD is taken to be a surrogate for actual damage, a practice validated in a separate study involving microscopic observation of damaged cell walls after various numbers of stress cycles (Gong and Smith, 2000). Transition cycles between stress levels had negligible influence on AD, and thus were presumed to have negligible effect on actual damage. Those cycles were disregarded when plotting results (Figure 6.6). The high-to-low load sequence resulted in larger AD than the low-to-high sequence. The AD curve leveled off for

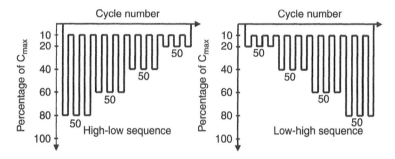

Figure 6.5 Simple stepped-load sequences: duty ratio $\tau = 0.50$.

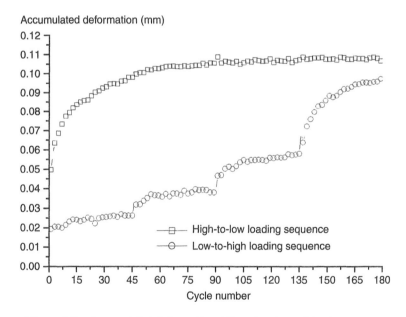

Figure 6.6 Accumulated deformation AD under stepped-load sequences.

---

[3] In the diagram C_{max} signifies static compressive strength, and duty ratio τ is the proportional period of residence at the peak stress level for the waveform.

the high-to-low sequence after the first block of stress cycles. Under the low-to-high sequence, *AD* increased progressively with any increase in the number of stress cycles and the number of blocks of stress cycles. Overall, the high-to-low load sequences was more detrimental to the wood than the low-to-high sequence. It is speculated that there is little accumulation of damage during stress cycles that have a peak level of stress lower than the maximum stress ever previous encountered. This contention is supported by results of cyclic creep tests of up to 35 days duration on nail embedment specimens (Whale, 1988). Although only proven for essentially compressive stress states, it is likely that the observed effect of load sequencing is widely applicable.

Based on cyclic bending tests on clear spruce specimens at loading frequencies of 0.1 and 1.0 Hz, Kohara and Okuyama (1992) showed that there is interaction between damage processes for cyclic fatigue and static fatigue effects. Those researchers found the interaction relationship to be:

$$\left(\frac{t}{T}\right) + \left(\frac{n}{N}\right)^{0.02} = 1.0 \tag{6.1}$$

where t is the total loading time to failure, n is the total number of load cycles to failure, T is the 'pure' static fatigue lifetime, and N is the 'pure' cycling fatigue. The equation describes points lying on the failure surface in a $t - n$ space. Table 6.3 gives predictions from Equation (6.1) based on LCF data for clear spruce loaded in compression parallel to grain. A difficulty is that it is not possible to actually establish the pure fatigue life of a rheological material. Therefore, N must be approximated as fatigue life under a low duty ratio, in this case $\tau = 0.05$. In Table 6.3 the term Pn is ratio of the 'load cycling effect' $(n/N)^{0.02}$ to the damage index SUM. It can be seen that the Kohara and Okuyama model fits the data quite well as SUM $\approx$ 1.0 for all duty ratios. This is not actually surprising, because the strength of clear wood loaded in bending depends upon the behaviour of material on the compression side of the neutral axis. It is clear that both cyclic fatigue and static fatigue can contribute significantly to damage processes even for quite large τ at relatively high loading frequencies. In general the extent of the load cycling effect will depend upon the uploading rate (waveform), which usually is directly proportional to the loading frequency. Clearly, it can be quite non-conservative to employ simplistic parameters such as cumulative time under load as the basis of life predictions. Unfortunately, as already discussed, this is what is done currently in structural design.

A common supposition is that cyclic creep strain for R-ratios greater than 0.0 (non-reversed stress) will not exceed the creep strain observed under the peak stress applied during cycles. It is also supposed that cyclic creep strain becomes asymptotic to the creep strain as τ approaches 1.0. Figure 6.7 shows the influence of a high σ on both

Table 6.3 Application of nonlinear interaction model: Equation (6.1)

Duty ratio, τ	t/T	$(n/N)^{0.02}$	SUM	Pn
1.00	1.0	0.0	1.0	0.0
0.95	0.03838	0.96422	1.00260	0.962
0.50	0.10641	0.99680	1.10321	0.904
0.05	0.01249	1.0	1.01249	0.988

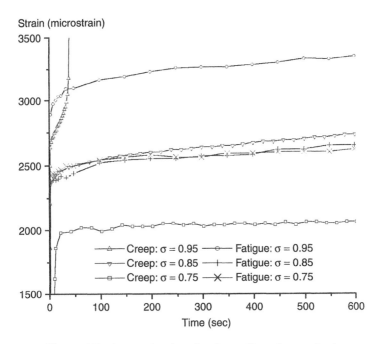

Figure 6.7   Accumulated strains for cyclic and creep load.

pure and cyclic creep strain for dry clear spruce loaded in compression parallel to grain. Tertiary creep was observed in creep specimens with 9 of 11 specimens failing during the test ($\sigma = 0.95$), with an average time to failure of 40 seconds. No specimens failed under cyclic load at that stress level. At σ of 0.75 the pure creep strain was much less than the cyclic creep strain, and at σ of 0.85 the pure creep strain and cyclic creep strain were approximately equal. Implication of the data is that there is some kind of threshold at σ of about 0.85. The well accepted notion that pure creep strain always bounds cyclic creep strain is invalid. This reflects that under pure creep damage depends mainly on the structural change in cell walls during uploading. Thus, at very high stress levels damage caused during uploading weakens material causing it to fail rapidly once the target σ has been reached. Damage due to cyclic stress is a function of σ and n. Bonfield *et al.* (1994), who studied HCF and creep behaviour of structural grade chipboard loaded in flexure observed pure creep specimens have larger strains than fatigue specimens. This conflicts with results of the study that underpins Figure 6.7. Possibly creep and failure mechanisms differ between wood and wood-particle based composites that contain glue.

The importance of stress ratio, R-ratio, on cyclic fatigue behaviour of wood has been established in a number of studies. Any increase in R-ratio corresponds to an increase in mean stress. For Southern pine and Douglas fir lumber loaded in flexure, Lewis (1962) showed that the higher the R-ratio, the lower was the fatigue strength for positive R-ratios. Tsai and Ansell (1990) studied the effect of the R-ratio on $\sigma - N$ curves for sliced African mahogany laminated veneer lumber loaded in flexure. Figure 6.8 summarises their results, and demonstrates that as the R-ratio approaches unity (static stress) the fatigue life increases. Fully reversed load results in the shortest fatigue life

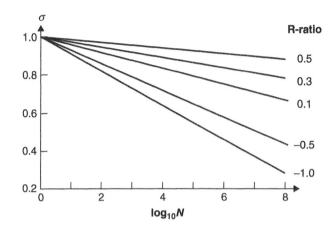

Figure 6.8   Influence of *R*-ratio on the fatigue strength of African mahogany LVL loaded in four-point flexure at a moisture content of 10% (based on data from Tsai and Ansell, 1990).

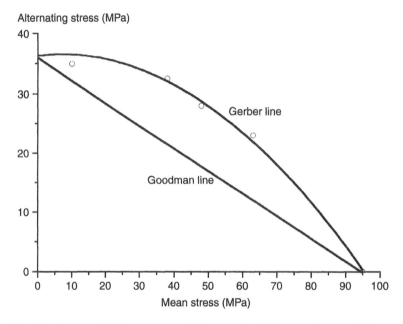

Figure 6.9   Goodman and Gerber constant life curves for African mahogany LVL at a life of 10^7 cycles [schematic] (based on Tsai and Ansell, 1990).

at a given σ. Constant life diagrams can be plotted as a derivative of such information, as shown in Figure 6.9. The straight-line relationship $\sigma_a = \sigma_e \left(1 - \frac{\sigma_m}{\sigma_u}\right)$ is known as a Goodman line, and the curved relationship $\sigma_a = \sigma_e \left(1 - \left(\frac{\sigma_m}{\sigma_u}\right)^2\right)$ as a Gerber line, where σ_a is the alternating stress, σ_e is the stress amplitude at R-ratio = 1.0, σ_m is the mean stress and σ_u is the ultimate strength. As can be seen, the Gerber line fits data much better than the Goodman line. The data and the Gerber relationship suggest that for negative values of R-ratio, the stress amplitude is a greater influence than the mean stress on N. For increasingly negative R-ratio, the compressive component of the fatigue load cycle plays a more important role in damage accumulation than the tensile component. This is because LVL, like clear wood, is weaker in compression than tension, and the damage rate is greater in compression than tension. Without confirmation based on testing, it should not be presumed that the Gerber or another line represents other stress states or other wood products.

Bonfield and Ansell (1991) carried out axial load fatigue tests on African mahogany LVL with the maximum stress being about equal to the compressive strength at R-ratio $= -10.0$, -2.0, -1.0, 0.1 and 10.0. Again, results showed R-ratio $= -1.0$ to be the most severe situation. At other values of R-ratio, fatigue life became progressively longer until R-ratio $= 10.0$, at which the σ vs. N trend was almost horizontal. Fatigue lives measured in all-tension tests (R-ratio $= 0.1$) were greater than those in all-compression tests (R-ratio $= 10.0$).

Clearly, the R-ratio strongly influences fatigue life under repetitive load cycles, with an R-ratio $= -1.0$ being the worst case scenario in all recorded studies. For positive R-ratios (non-reversed loading), mean stress is the dominant factor determining fatigue life, while for negative R-ratios (reversed loading) stress amplitude is the dominant factor.

6.2.3   Residual mechanical properties

The residual strength of wood-based materials following cyclic load was first investigated by Kommers (1943a), who carried out displacement-controlled flexural tests on five-ply Sitka spruce plywood. Five thousand cycles of non-reversed load were applied at various σ, after which specimens were statically loaded until failure either with load in the same or opposite direction as during load cycling. No measurable reduction in strength was observed when residual capacity was measured in the 'same loading direction', even if the cyclic peak σ was 0.85. In fact, a small increase in strength relative to that of matched statically loaded material was observed. Residual capacity in the 'opposite loading direction' was considerably less than matched static strength, because tension activated damage in previously compressed material. More recent studies have shown that in a qualitative sense Kommers results apply to clear wood load parallel to grain (Kellogg, 1960; Rose, 1965; Dobraszczyk, 1983). Gong (2000) performed non-reversed cyclic tests on small clear spruce specimens in compression parallel to grain at $\sigma = 0.95$, and found the average residual strength of fatigued specimens is 4% higher than that of end-matched specimens subject to static loading.

Various investigations have provided indications of how stiffness of wood parallel to grain changes under cyclic load when σ exceeds the proportional limit. Change is expressed in terms of variation in the residual tangent modulus (often called the effective elastic modulus) due to stress cycling. Effective modulus includes effects of delayed elastic and viscous components of deformation. Kommers (1943b) investigated the effect of ten cycles of bending or compressive stress on the effective modulus of clear Sitka spruce and Douglas fir. The peak σ was approximately 0.95, and was therefore well above the proportional limit. Effective modulus decreased with each cycle with the greatest incremental decrease occurring as a result of the first cycle. Clorius *et al.* (1996) discovered that the stiffness of wood subjected to parallel to grain compression tends to increase in the first cycle, and then decrease in subsequent cycles. This result conflicts with others, and could be explained by local densification of specimens. Kellogg (1960) studied the effect of repetitive loading on tensile properties of nine different species of air-dry softwood and hardwood. Specimens were loaded for 100 cycles at σ between 0.30 and 0.80, before being loaded to failure. There was no significant change in effective modulus after 100 cycles. Dobraszczyk (1983) also reported no significant change in effective modulus during cyclic tension tests. However, there was a quite marked decrease during cyclic flexure tests, because of sensitivity of material on the compression side of the neutral axis to load cycling. The rate of decrease in the effective modulus in flexure or compression increases with any decrease in the loading frequency because that increases creep strain. Clearly, loading mode and σ both strongly influence the extent of any change in residual stiffness of wood loaded parallel to grain. It is probable that changes in effective modulus reflect development of fatigue induced damage at the cellular level, but there has been no direct observational proof of this.

6.2.4 Density effect

Sekhar and Shukla (1979) studied the effect of density on flexural fatigue in wood, with tests being conducted on 15 wood species representing air-dry relative densities ranging from 0.32–0.73, using a loading frequency of 23.7 Hz. Based on that, and data from their earlier research, they showed that fatigue life increases with any increase in relative density and decreases with any increase of stress level (Figure 6.10). The influence of relative density at high stress levels (0.46–0.60) is not as strong as at low stress levels (0.30–0.45). Sieminski (1960) also found a linear relationship between relative density and fatigue strength from flexural fatigue tests on Scots pine.

6.2.5 Massive wood

Discussion here focuses on sawn solid wood products (lumber and heavy timbers) and substitute composite materials manufactured using waterproof structural adhesives (glued-laminated-timber (glulam), and Laminated Veneer Lumber (LVL)). As discussed in Chapters 2 and 3, structural wood materials can exhibit gross irregularities. Most irregularities in lumber are due to inherent features of tree stems (e.g. knots, spiral grain), while others are produced as by-products of manufacturing processes. Permitted extent of irregularities varies between grades of lumber. Composites, although intended to homogenise properties, also contain irregularities, e.g. intra-laminate or exo-laminate joints in glulam or LVL. Because of irregularities, which are usually stress-raising features, it is possible for fatigue behaviour of massive wood members to differ from that of clear wood of the same species.

Static strength of massive wood exhibits a relationship between specimen volume and apparent strength. Although there are exceptions, apparent strength tends to reduce proportional to any increase in volume (Madsen, 1992). Reasons for so-called volume or size effects are complex. Factors involved include how absolute member/laminate

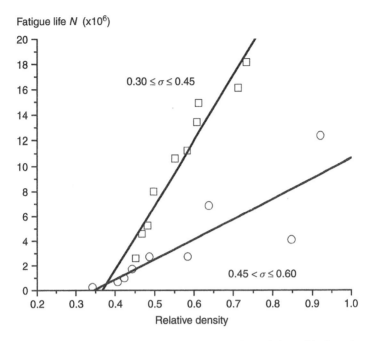

Figure 6.10 Effect of relative density and stress amplitude on fatigue life (based on data from Sekhar and Shukla, 1979).

sizes compares to sizes of parent trees, sawing pattern used during log conversion, 'stress grading' technique used to control magnitude of strength reducing features, and type and frequency of any 'finger joints'. It is expected that shape, size and material quality of massive wood components will influence fatigue properties. Knowledge cannot be simply borrowed from studies at the macro scale (clear wood), and it can be anticipated that fatigue behaviour of massive wood will be application specific to some extent.

Effects of cyclic load

Deviation of the grain from the member axis direction is a primary cause of lumber's reduced strength in comparison to clear wood under axial and bending forces (FPL, 1989; Madsen, 1992). Grain deviations can manifest themselves as local disturbances around features such as knots and gross disturbances that affect the whole cross-section. The latter can be in the form of 'general slope-of-grain' or a localised 'cross-grain' condition. Lewis (1962) investigated the effect of general slope-of-grain on fatigue strength of Southern pine and Douglas fir. Straight-grained air-dry lumber was found to have slightly higher fatigue strength than lumber with a general slope-of-grain condition. For example, fatigue strength in flexure at two million cycles was estimated to be 60% of the static strength for straight grained lumber of either species, but only 50% and 55% of static strength at a 1 : 12 slope-of-grain for Southern pine and Douglas fir respectively. Sekhar and Shukla (1979) concluded that the use of wood with a high slope of grain should be avoided under cyclic stress conditions based on a study of the effect of grain angle on reversed flexural fatigue properties of Deodar cedar at a loading frequency of 23.7 Hz. They suggested material loaded cyclically should have a general slope-of-grain less than ten degrees (1 : 5.7). Hansen (1991) studied the effect that slope of grain has on fatigue properties of laminated wood beams loaded in four-point flexure at a loading frequency of 10 Hz. He observed marked decrease in fatigue properties due to any increase in the grain angle in the range 0–12 degrees (1 : 4.7). Lewis (1962) examined the effect of man-made checks on Southern pine and Douglas fir. Specimens were loaded in flexure at a frequency of 8.33 Hz, R-ratio = 0.1. Both static and fatigue properties were greatly reduced by checks, but when fatigue strength was expressed as a percentage of static compressive strength there was only a small negative effect due to checks.

Lewis (1962) compared fatigue properties of green and air-dry Southern pine and Douglas fir. Green specimens with straight grain loaded in flexure had fatigue strengths of 50% of static strength at two million cycles for Southern pine and 55% for Douglas fir. The results for air-dry specimens were slightly higher at 60% of static strength for both species. Therefore, for lumber it appears that there is mild sensitivity of fatigue strength to moisture content. This is consistent with findings from static strength tests on lumber, which have shown tensile and bending strengths of low quality lumber to be almost completely insensitive to the moisture condition (Madsen, 1992; Zhou and Smith, 1991). By contrast, studies on flexural fatigue in African mahogany LVL showed a detrimental effect of increased moisture content on fatigue life, based on moisture content $m = 5\%$, 11% and 35% (Figure 6.11) (Tsai and Ansell, 1990). The implication is that the rate at which fatigue damage accumulates increases with any increase in m for massive LVL members, between typical indoor and fibre saturation conditions. The two studies just mentioned demonstrate that the influence of moisture content on fatigue properties of massive wood relates to the quality of the material, with sensitivity diminishing with lower quality.

Structural members are quite often used in marine applications, and Freas and Warren (1959) studied how saltwater immersion effected flexural cyclic fatigue strength of

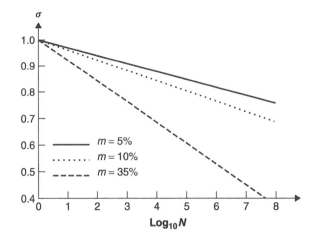

Figure 6.11   Influence of moisture content on fatigue life of African mahogany LVL loaded in flexure, R-ratio = 0.1 (based on data from Tsai and Ansell, 1990).

laminated white oak members. Specimens tested dry or salt-water immersed survived nine million cycles at load at a peak σ of 0.5. Although there was no appreciable effect of salt on residual strength or stiffness relative to matched new material, wetting due to saltwater immersion itself reduced properties by about 25%.

The effects of 'coal-tar creosote' preservative on static and fatigue properties of Southern pine and Douglas fir lumber has been studied (Lewis, 1962). Treatment reduced both static and fatigue strength. This is attributable to damage in cells caused by the treatment process because creosote does not chemically modify wood substance. Other chemicals applied as fire retardant or preservative treatments can modify wood substance either through immediate or long-term processes. It is to be expected that chemical formulation, the treatment process and service environment (temperature and moisture conditions) will all influence effects of various treatments on wood properties. Modern chemical treatment substances and processes are proprietary, and it is prudent to obtain information from manufacturers on how their products affect mechanical properties, including fatigue characteristics.

An important question with regard to relatively fast cyclic loading is whether previous load excursions have measurable influence on residual strength of structural wood members. Studies on 'proof grading' of lumber give insight into this. Proof grading is a process in which pieces of lumber are passed through machines that apply a certain proof level of stress as a means of qualifying the material for structural use. The method is intentionally destructive for pieces having initial strength less than the proof stress. Nominally it is presumed that material that survives the process has residual strength at least equal to the proof stress, but that is not always the case. Proof grading can cause damage, with pieces that have residual strength less than the proof stress termed rogues. Leicester (1990) considered situations where 'small dimension' lumber is proof graded in bending using single-pass or double-pass loading systems. Single-pass grading systems only load one edge of lumber in tension. The tension edge can be 'selected' at random (unbiased selection), or deliberately with the intent of loading the weakest orientation (biased selection). Double-pass systems subject lumber to reversed loading, in which case both edges are loaded sequentially in tension then compression or compression then tension. Assuming that the tension face determines the strength of any piece of lumber, which is usually the case, there are two causes of rogues with single-pass systems. First, the edge loaded in tension may not have contained a critical strength determining feature because as it was located on the other edge. Secondly, the residual strength can be compromised because of damage induced by the proof stress. Leicester terms this 'single cycle' fatigue, with the R-ratio = 0.0. In double-pass grading systems, rogues result only from single cycle fatigue (R-ratio = −1.0). The extent of single-cycle fatigue was estimated based on a laboratory simulation using non-reversed (single-pass) and reversed (double-pass) bending load. It was found that significant levels of damage can be caused by either single-pass or double-pass proof grading. For 35 × 90 mm Radiata pine lumber measurable damage occurred in between 3% and 25% of specimens, with the incidence of rogues depending upon the load arrangement (3-point versus 4-point load arrangements), diameter of the loading head, and the type of defect(s) in the lumber. For combinations of test variables studied, the average residual strength of rogues was between 83% and 97% of the proof load level. Weakest rogues had residual strengths ranging from 57–96% of the proof load level. The study by Leicester clearly demonstrates that survival of any load event does not reliably imply anything about residual strength of wood components. It also lends support to the notion that for random load pulses of short duration most damage is done by the most severe load pulse (or during the first visit during repetitive cycles).

Effects of sustained load

Over the last 20 years there have been many studies on static fatigue in lumber members of sizes realistic for end use applications in light-frame construction. Such tests gather engineering design level information (Foschi *et al.*, 1989), with many reports being available (e.g. Glos, 1986; Karacabeyli, 1988; Karacabeyli and Soltis, 1991; Madsen, 1992; Hoffmeyer, 2003). Although useful for the intended purpose, data is interpreted empirically and adds little to scientific understanding. Tests are repeated on a product by product basis and apply to a particular loading and moisture conditions (Figure 6.12).[4] An interesting finding of such work is that softwood lumber subjected to tensile or compressive stress parallel to grain accumulates damage more rapidly under sustained load than does matched material subject to flexural load of comparable stress level. This is attributed to there being a greater chance that strength reducing features such as grain perturbation around knots will be critically loaded, and because there is no possibility of stress redistribution under axial load conditions.

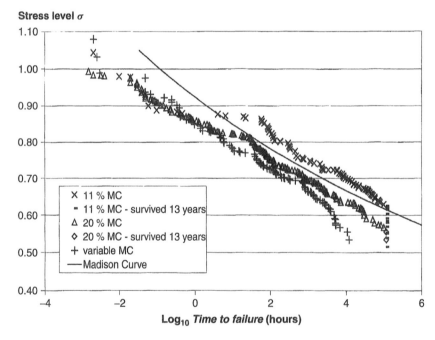

Figure 6.12 Stress level versus Log_{10} Time to failure for 50×100 mm beams of Norway spruce lumber subjected to bending at MC (moisture content) $= 11\%$, MC $= 20\%$ and MC varying between 11% and 20% (adapted from Hoffmeyer, 2003).

---

[4] For comparison, the diagram shows the so-called Madison curve (Woods, 1951) that applies to clear Douglas fir loaded in bending.

Sustained load tests on lumber indicate very substantial reductions in residual strength of softwood lumber following one or three years sustained loading at σ of about 0.75 and 0.60, respectively (Karacabeyli, 1988). Practical and economic demands have usually militated against collection of data for stress levels lower than those resulting in time to failure greater than about four years, but there are a few exceptions (Gerhards, 2000; Hoffmeyer, 2003). There is no conclusive experimental proof, and hence no agreement on existence, or absence, of a threshold σ_0 below which damage will never accumulate. Existence of such a threshold has been postulated and incorporated into some types of empirical damage model (Barrett and Foschi, 1978a, 1978b; Foschi and Yao, 1986), as is discussed in Chapter 8. Values of σ_0 for softwood lumber have been estimated to lie in a range of about 0.35–0.55, depending on wood species and quality (Foschi *et al.*, 1989), with the value being greater the higher the quality. However, fitting of thresholds involves adoption of damage functions that presuppose they exist beyond the temporal range of data. All that is actually known is that if ascending levels of constant stress are applied to matched groups of specimens there is an inverse relationship between the level of stress and average time to failure. Despite any material matching process, strengths of individual specimens within a set vary and actual σ can only be approximated. There is typically quite large variability in T, with values being positively skewed (Figure 6.13a). There is an approximately linear relationship between the level of constant stress σ and $\log_{10} T$ (Figure 6.13b).

Various researchers have employed the so-called equal rank assumption as a means of estimating the actual σ for each specimen and as a means of smoothing static fatigue data. The technique also has the 'benefit' of effectively extending the range of stress considered at either end of the range, e.g. results shown in Figure 6.12. Under the assumption it is presumed that the ordering of observed times to failure under a certain level of apparent stress is positively correlated with cumulative frequency in the static strength distribution (Madsen, 1992). Thus, the control strength for the ith specimen to fail under static load is deduced from the static strength distribution.[5] With this mapping procedure it is possible to plot relatively smooth σ versus $\log_{10} T$ trends. Employing the equal rank assumption gives coherence to the data. It can lead to a definite negative curvature in the $\sigma - \log_{10} T$ relationship at about the lowest deduced σ used in tests (Figure 6.13c). As shown in the figure, the extrapolation of the data and the estimated threshold σ_0 reflect fitting of a damage accumulation model of the 'Barrett and Foschi type' (see Section 8.2). Approximately speaking, it is reasonable to compare the mapped trend with a correlation trend of the type in Figure 6.13b. It is apparent that shapes of 'duration of load effect on strength' curves are highly dependent on data interpretation and analysis techniques. As is discussed in Section 8.4, evidence in support of a σ_0 is dubious and the notion appears conceptually flawed. Analysts should exercise great care before accepting specific interpretations of data, and be cognisant of links between this and subsequent use of curves in, for example, structural reliability analysis.

Some researchers have attempted to measure directly whether loads compromise residual capacities of structural wood members that survive in service. The approach taken is to sample material from decommissioned structures and measure residual

---

[5] If the distribution form is unknown, the cumulative frequency associated with the ith specimen to fail under static load can be taken as $i/(n_r + 1)$, where n_r is the number of replicates.

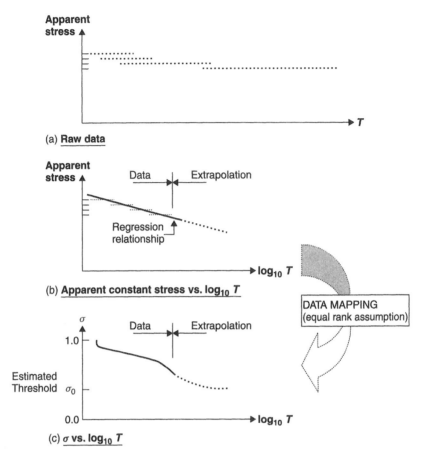

Figure 6.13   Trend and interpretation of data from constant apparent stress (static fatigue) tests on massive wood members (schematic).

strength (Kuipers, 1986; Falk *et al.*, 1999, 2000). For example, Kuipers (1986) sampled softwood glued-laminated-timber members that had been in service for 52 years in the RAI-hall (was at that time the oldest glulam structure in The Netherlands); large softwood timbers that had been in service in a locomotive shed in The Netherlands for between 100 and 120 years; and tropical hardwood ship bulkhead planks that had been in service for 15 years. Residual strength properties were greater than those of new material of similar density. Implication of such findings is that there was no dramatic duration of loading effect on strength of sampled material. A reason could be that there was a threshold stress level, and that it was never exceeded. If so, the structural engineering level conclusion could be drawn that it is only necessary to account for effects of the peak lifetime load events during design. In turn this implies need for only limited knowledge of static fatigue under moderate or low σ, as peak lifetime load events tend to be of limited duration. However, there are various difficulties associated with interpretation of such anecdotal data, not least being lack of knowledge of the original strength of the material and levels of stress experienced by members over their service life. Moisture movements, especially those associated with drying out or seasonal variation in climate, can lead to increased splitting and checking of wood members. This does not necessarily adversely influence residual strength of members under axial or flexural load (Falk *et al.*, 2000). However, for situations where shear or tension perpendicular to grain properties control member behaviour, residual capacities of members could be substantially impaired in service.

6.2.6   *Wood-based panel products*

Use of plywood in aircraft prompted study of its HCF performance during the 1940s. Kommers (1943a) measured residual strength of five-ply Sitka spruce plywood following 5000 cycles of non-reversed flexural load, using various σ and a loading frequency of 30 Hz. Specimens were cut so that face veneers would be stresses parallel to grain. For a σ as high as 0.85, he found no reduction of residual strength if specimens were flexed in the same direction as during load cycling. This did not mean there was no damage, as considerable reduction was observed in residual strength when specimens were flexed in the opposite direction to that during load cycling. Damage due to cyclic loading was restricted to the compression side of specimens because the tensile strength of face veneers was much greater than their compressive strength. The damage was of a type that only propagated unstably if stress was reversed after its creation. In essence, the plywood behaved like clear wood loaded in flexure (see Section 6.3.2). In a separate study using five-ply plywood with Sitka spruce, yellow birch and yellow poplar veneers, fatigue strength of plywood was found to be about 27% of the static strength after 50 million cycles of reversed flexure (Kommers, 1943c). Clearly, residual strength depends upon the R-ratio.

Cyclic fatigue performance of panels with wood particles has been investigated in various studies. Bao and Eckelman (1995) carried out flexural fatigue tests on Medium Density Fibreboard (MDF), Oriented Strandboard (OSB) and particleboard loaded in flexure at a loading frequency of 0.33 Hz. Peak σ ranged from 0.3–0.7, and tests continued until one million cycles or premature failure. Fatigue life reached an extremely large number ($>10^6$) when the stress level was 0.3 or less, leading to the deduction that the fatigue limit is about 30% of static strength. Specimens were loaded in planar-shear and inter-laminar shear (shear through the thickness). Peak σ ranged from 0.38–0.90 for tempered hardboard, and from 0.45–0.90 for particleboard. The loading frequency was 15 Hz with an R-ratio = 0.1. It has been shown that $\sigma - N$ curves for tempered hardboard and various types of construction grade particleboard made with synthetic adhesive are similar to those for solid wood loaded in tension parallel to grain or glued wood shear specimens (McNatt, 1970; McNatt and Werren, 1976).

Bonfield *et al.* (1994) carried out flexural cyclic fatigue tests on structural grade particleboard with relatively high adhesive content. Peak σ varied between 0.50 and 0.80, the loading frequency was about 5 Hz, the R-ratio was 0.01, and matched static fatigue tests were conducted with static load levels equal to the various peak stress levels in cyclic fatigue tests. Static fatigue specimens never fail before cyclic fatigue specimens loaded at the same σ. Until close to failure, pure creep strain nearly always was greater than cyclic creep strain, suggesting that damage mechanisms differ between static and cyclic fatigue. It was deduced that damage results mainly from the first load cycle and is proportional to the σ, which is consistent with observations on clear wood loaded parallel to grain (Gong, 2000).

Overall, although the data is quite sparse, qualitatively at least, fatigue behaviour of wood-based panel products is similar to that of clear wood even in the case of products such as OSB that incorporate wax and synthetic resin. The fatigue limit for wood based panel products is about 30% of static strength.

6.3 Failure Mechanisms in Clear Wood

In this section, consideration is given to fatigue induced failure mechanisms as they appear at the macro scale (clear wood). In clear wood fatigue occurs by accumulation processes with small cracks initiating at many points throughout a specimen before coalescing to produce a failure surface(s). Because wood is a natural aligned-fibre reinforced composite, it is more sensitive to compressive stress than tensile stress in the grain (fibre) direction, and marked fatigue sensitivity to compression parallel to grain is to be expected. Conversely, alignment of fibres means that wood is extremely sensitive to tension when stressed normal (perpendicular) to grain.

Most fatigue studies have employed flexural loading that develops stress parallel to the grain. However as this involves a complex mix of compression, tension and shear effects, results of flexural studies cannot be interpreted unambiguously. Studies on fatigue due to pure stress states (axial or shear stress) are preferable for elucidation of mechanisms. Another important consideration is that postmortem examination of specimens is not particularly helpful because failure surfaces are usually indistinguishable for static, static fatigue and cyclic fatigue stress. Only experiments specifically designed to enable real-time observation of damage at the microscopic level elucidate damage processes unambiguously. Experimental evidence relates to damage processes under nominal rather than true stress conditions. Because of its anisotropic and heterogeneous nature, whatever the nominal stress condition, damage initiates due to tension and shear at sub-macro (nano or micro) scales. The rest of the section discusses fatigue processes at macro and sub-macro scales under different types of stress.

6.3.1 Axial load parallel to grain

Possible fatigue mechanism(s) in axial tension include slippage between and within cell wall layers, cracking in secondary wall layers and splitting along the grain (Tsai and Ansell, 1990). Spruce fatigued in tension parallel to grain has been shown to accumulate minute cracks with each load cycle until damage reaches a critical level and failure occurs (Okuyama *et al.*, 1984). Figure 6.14 shows the failure of Jack pine due to cyclic stress parallel to grain as observed in a SEM. It was found that slippage occurs along the boundary between two growth rings, and tracheids are pulled out and broken normal to the axis of those cells. Fatigue mechanisms reflect specific structure of different species, and it is almost impossible to generalise about them.

Clorius *et al.* (1996) studied the effect of loading frequency on the fatigue behaviour of spruce subjected to square wave compressive stress parallel to grain. The σ_{min} was 0.80 times the static strength and loading frequencies were 0.01, 0.1, 1 and 10 Hz. A 'fine-meshed' failure pattern was observed at low frequencies, while shearing failures

se Tensile failure in Jack pine 900 μm

Figure 6.14   Failure of Jack pine (*Pinus banksiana*) due to cyclic tensile stress parallel to grain.

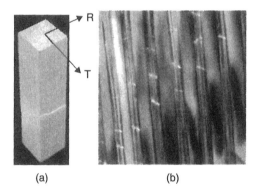

(a)   (b)

Figure 6.15   Failures under compressive stress parallel to grain. (a) Gross shear band in black spruce (*Picea mariana*), (b) Kinks in tracheid walls of black spruce (200X).

occurred at high frequencies. Gong and Smith (2000) studied the formation and development of kinks of spruce specimens under compressive static, creep and fatigue loads (Figure 6.15). They found that kinks are more uniformly distributed across tracheids in the R-L plane under creep loading than under static loading. Most kinks exist in the latewood tracheids. However, in fatigue tests, kinks are generated among the latewood tracheids close to the boundary between two growth rings, and then develop from latewood to earlywood tracheids resulting in buckling of the earlywood tracheids. Kinks accumulate quickly as the number of load cycles is increased, and many new kinks are initiated, which agrees with the finding by Tsai and Ansell (1990), who studied specimens loaded in flexure.

Fracture topography of fatigued African mahogany LVL was studied using scanning electronic microscopy by Bonfield and Ansell (1991). They found that in tension there is evidence of 'bunches' of cells pulled out and regions of clean cross cell fracture, with an absence of crushed cell ends, R-ratio = 0.1. Under compression, the failure path was demarcated by macro shear bands that lay at approximately 45° relative to the axis on tangential faces. At shear bands, specimen surfaces were fairly smooth on the radial plane and slightly ridged on the tangential plane of wood veneers. These observations suggest that although LVL is often available in quite large dimensions its behaviour is similar to that of clear wood.

Smith and Gong (2002) studied the low cycle fatigue behaviour of dry knotty spruce lumber loaded in compression parallel to grain. They found that failure of lumber specimens containing knots results mainly from macroscopic creases that generate from microscopic kinks in cell walls. Such creases occur in clear wood adjacent to knots, after consolidation of knots. Although ultimately cracking is observed, it is a consequence of rather than an initiator of the failure mechanism. They suggest that concepts and models for predicting compressive behaviour of knotty lumber in the parallel to grain direction should be based on development of creases as in clear wood.

6.3.2 Flexure parallel to grain

In flexure, damage on the compression side of the neutral axis in the form of kinks, or creases, does not necessarily lead to catastrophic failure. Tsai and Ansell (1990) report optical observations of microstructural flexural fatigue damage due to non-reversed load. Sitka spruce specimens were fatigued at a peak σ of 0.75, R-ratio = 0.1. Manifestations of damage were creases on the compression face and longitudinal cracks on the tension face. The compression damage was observed by polarised light microscopy in sections microtomed after 100, 500, 10^3, 10^4 or 10^5 cycles. Damage was progressive

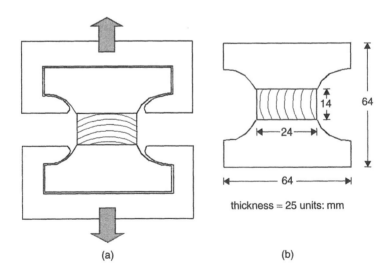

(a)   (b)

Figure 6.16 (a) Experimental set-up and (b) dimensions of specimens used in tensile fatigue tests perpendicular to grain: (a) radial loading and (b) tangential loading.

and was first observed in the form of fine microscopic creases after 500 cycles. After higher numbers of load cycles, the kinks spread in the depth of the beam towards the neutral plane, and along cell walls in the longitudinal direction. On the tension face, a first small crack was followed by number of others. Once a crack formed, it grew longitudinally to coalesce with other cracks to form a splinter. Extensive cracking was observed during the last third of specimen life. When flexural load was reversed, cracks in specimens are very different from those formed under non-reversed load. Under reversed load, splinters were observed and usually had blunt tips, indicating

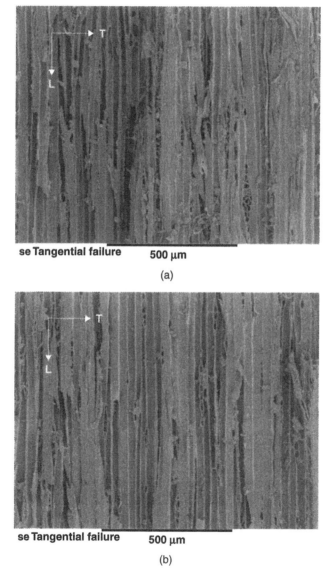

se Tangential failure   500 µm

(a)

se Tangential failure   500 µm

(b)

Figure 6.17 Tangential fracture surfaces of Jack pine (tensile stress in radial direction): (a) under static load, and (b) under cyclic load.

that tensile failure occurred after formation of compression creases. Formation of one crack rapidly leads to others on both faces of a specimen and then catastrophic failure (Tsai and Ansell, 1990).

6.3.3   Axial stress perpendicular to grain

Exploratory studies have been undertaken at the University of New Brunswick into fatigue failure mechanisms in tension perpendicular to grain using laminated butterfly specimens (Figure 6.16). Upper and lower parts of specimens were Maple. Their purpose was to permit gripping of the specimens and to avoid any damage to the central portion as can occur during manufacture of necked specimens. The central portion was Jack pine and had growth rings oriented to produce radial or

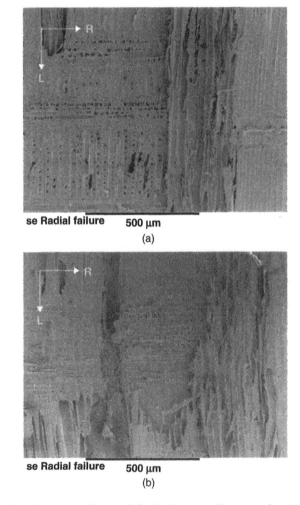

(a)

(b)

Figure 6.18   Radial fracture surfaces of Jack pine (tensile stress in tangential direction): (a) under static load, and (b) under cyclic load.

tangential stress. Failure always occurred in the relatively weak Jack pine. The load frequency was 0.5 Hz, the waveform triangular, and R-ratio = 0.11. It was found that fracture surfaces due to radial loading are much rougher in fatigued specimens than in corresponding static specimens. In most cases, static load fracture surfaces exhibit transwall fracture coinciding with rays (separation of ray parenchyma), especially in earlywood (Figure 6.17a). However, for fatigue load, most specimens exhibit inter-cell/intrawall fracture, with many tracheids being pulled out in both earlywood and latewood (Figure 6.17b). This suggests occurrence of new damaging loci as cyclic loading progresses. Under tangential loading, the fracture surfaces are similar for both static specimens and fatigued specimens. Interestingly, transwall fracture and inter-cell/intrawall fracture are more uniformly distributed in fatigued specimens than in static specimens (Figure 6.18).

Failure mechanisms of wood subjected to fatigue loading in compression perpendicular to grain have also been studied at the University of New Brunswick. Three layer laminated specimens were employed for both radial and tangential loading (Figure 6.19). The top and bottom pieces were maple and transferred stress uniformly to the core pieces of Aspen or Jack pine. Preliminary results reveal that failure under radial fatigue load is preferentially located in earlywood at the beginning of a single growth ring for both Aspen and Jack pine. This is similar to what happens under static load. Several layers of fibres or tracheids are squeezed and collapsed, finally resulting in a failure band. In Aspen, cavities of vessels in the failure bands are collapsed resulting in large deformation (Figure 6.20). Under tangential load, a failure band forms as the result of several collapsed fibres or tracheids. The band seeks the path of least resistance which is off-axis relative to the rays. Failure is very localised in earlywood

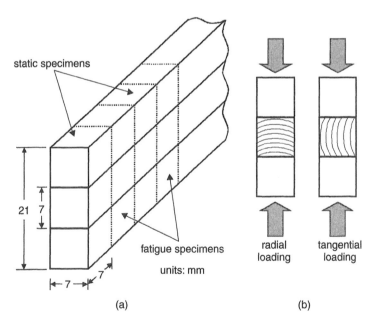

Figure 6.19 (a) Dimensions and (b) experimental set-up and loading direction used in compression perpendicular to grain fatigue tests.

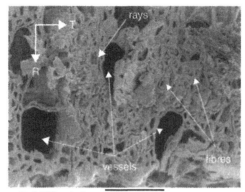

se Radial failure in Aspen 100 μm

Figure 6.20   Collapse of fibres and deformation of vessels in Aspen under cyclic compressive stress in the tangential direction.

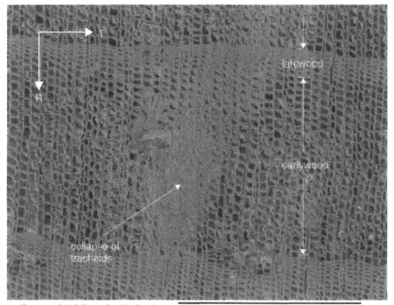

se Tangential failure in Jack pine   700 μm

Figure 6.21   Failure of Jack pine under tangential compressive fatigue load.

of Jack pine (Figure 6.21). By contrast, the failure pattern is quite uniform in Aspen (Figure 6.22).

6.3.4   *Shear*

Preliminary studies at the University of New Brunswick have investigated fatigue failure in shear parallel to grain using butterfly specimens (Figure 6.23). Spruce was the

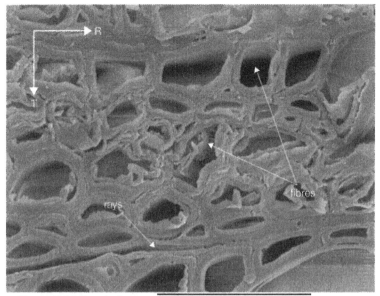

se **Tangential failure in Aspen   50 μm**

Figure 6.22   Failure of Aspen under tangential compressive fatigue load.

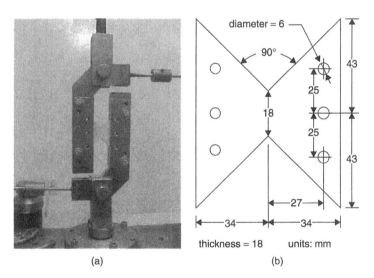

(a)   (b)

Figure 6.23   (a) Experimental set-up and (b) dimensions of specimens used in shear parallel
to grain fatigue tests.

test species with failures induced mainly in the radial plane. Results reveal that failures follow a mixture of transwall and intercell/intrawall fracture patterns under static
loads, however a transwall fracture pattern is favoured under cyclic load conditions
(Figure 6.24). This suggests that fatigue loads may create more damage loci as loading
proceeds, resulting in more breakage of tracheids than under static load.

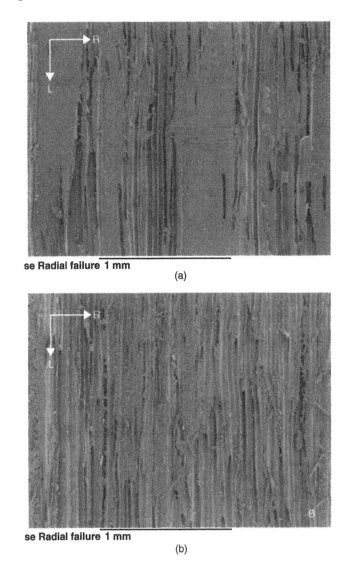

se Radial failure 1 mm

(a)

se Radial failure 1 mm

(b)

Figure 6.24 Radial fracture surfaces of spruce loaded in shear parallel to grain: (a) under static load, and (b) under cyclic load.

6.3.5 Summary

As illustrated above, fracture mechanisms are quite complex for both static and fatigue loading. Differences between fracture surfaces are however often sufficiently different to give insights into mechanism by which they developed. *In situ* observation of failure processes of wood using environmental electronic scanning microscopy (ESEM) would permit better appreciation of mechanisms, but to date no such observations exist. Under fatigue load there is often evidence that the number of damage loci increases as load cycling progresses. This conflicts with suggestions in the literature that static and fatigue load induced fracture surfaces cannot be distinguished. More correctly, distinguishing them could be said to be very difficult.

6.4 References

Ansell, M.P. (1995) 'Fatigue design for timber and wood-based materials', Lecture E22. In: Timber Engineering Step: Design — details and structural systems, Ed. Blass, H.J., Aune, P., Choo, B.S., Gorlacher, R., Griffith, D.R., Hilson, B.O., Racher, P. and Steck, G., Centrum Hout, Almere, The Netherlands: E22/1–E22/8.

ASCE Committee on Wood (1975) 'Wood structures: A design guide and commentary', American Society of Civil Engineers, New York, NY, USA.

ASTM (American Society for Testing and Materials) (1993) 'Standard definition of terms relating to fatigue', In: Book of ASTM Standards, Vol. 03.01, ASTM, Philadelphia, PA, USA.

Bao, Z.Z. and Eckelman, C.A. (1995) 'Fatigue life and design stresses for wood composites used in furniture', *Forest Products Journal*, **45**(7/8): 59–63.

Barrett, J.D. and Foschi, R.O. (1978a) 'Duration of load and failure probability in wood. Part I. Modelling creep rupture', *Canadian Journal of Civil Engineering*, **5**(4): 505–514.

Barrett, J.D. and Foschi, R.O. (1978b) 'Duration of load and failure probability in wood. Part II. Constant, ramp and cyclic loading', *Canadian Journal of Civil Engineering*, **5**(4): 515–532.

Bonfield, P.W. and Ansell, M.P. (1991) 'Fatigue properties of wood in tension, compression and shear', *Journal of Material Science*, **26**: 4765–4773.

Bonfield, P.W., Hacker,C.L., Ansell, M.P. and Dinwoodie, J.M. (1994) 'Fatigue and creep of chipboard: Part 1. Fatigue at R = 0.01', *Wood Science and Technology*, **20**: 423–435.

Buffon, G.L. Le Clerk, Comp de (1740) 'Experiences sur la force du bois', Histoire et Memoires, Paris L'Academie Royale des Sciences: 453–467.

Clorius, C.O., Pedersen, M.U., Hoffmeyer, P. and Damkilde, L. (1996) 'Fatigue damage in wood', *Proceedings of International COST508 Wood Mechanics Conference*, Office of Publications of the European Communities, Luxembourg: 229–242.

Daneff, G. (1997) 'Response of bolted connections to pseudodynamic (cyclic) loading', MScFE thesis, University of New Brunswick, Fredericton, NB, Canada.

Dobraszczyk, B. (1983) 'An investigation into the fracture and fatigue behaviour of wood', Ph.D. Thesis, University of Bath, Bath, UK.

Falk, R.H., Green, D.W. and Lantz, S.F. (1999) 'Evaluation of lumber recycled from an industrial military building', *Forest Products Journal*, **49**(7/8): 49–55.

Falk, R.H., Green, D.W. and Lantz, S.F. (2000) 'Engineering evaluation of 55-year-old timber columns recycled from an industrial military building', *Forest Products Journal*, **49**(7/8): 49–55.

FPL (Forest Products Laboratory) (1989) 'Handbook of wood and wood-based materials for engineers, architects, and builders', Forest Service, U.S. Department of Agriculture. Hemisphere Publishing Corporation, New York, NY, USA.

Foschi, R.O., Folz, B.R. and Yao, F.Z. (1989) 'Reliability-based design of wood structures', Structural Research Series Report No. 34, University of British Columbia, Vancouver, BC, Canada.

Foschi, R.O. and Yao, Z.C. (1986) 'Another look at three duration of load models', Paper 19-9-1, Volume II, *Proceedings of CIB-Working Commission 18: Timber Structures, Meeting 19, International Council for Research and Innovation in Building and Construction*, Rotterdam, The Netherlands.

Freas, A.D. and Warren, F. (1959) 'Effect of repeated loading and salt-water immersion on flexural properties of laminated white oak', *Forest Products Journal*, **9**: 100–103.

Gerhards, C.C. (2000) 'Bending creep and load duration of Douglas-fir 2 by 4s under constant load for up to 12-plus years', *Wood and Fiber Science*, **32**(4): 489–501.

Glos, P. (1986) 'Creep and lifetime of timber loaded in tension and compression', Paper 19-9-6, Volume II, *Proceedings of CIB-Working Commission 18: Timber Structures, Meeting 19*, International Council for Research and Innovation in Building and Construction, Rotterdam, The Netherlands.

Gong, M. (2000) 'Failure of spruce under compressive low-cycle fatigue loading parallel to grain', PhD thesis, University of New Brunswick, Fredericton, NB, Canada.

Gong, M. and Smith, I. (1999) 'Low-cycle fatigue behaviour of softwood in compression parallel to grain', *Proceedings Pacific Timber Engineering Conference, Forest Research Bulletin 212*, Research Institute Limited, Rotorua, NZ: 3.437–442.

Gong, M. and Smith, I. (2000) 'Failure mechanism of softwood under low-cycle fatigue load in compression parallel to grain', *Proceedings World Conference on Timber Engineering*, University of British Columbia, Vancouver, BC, Canada (on CD).

Hansen, L.P. (1991) 'Experimental investigation of fatigue properties of laminated wood beams', *Proceedings of International Timber Engineering Conference*, Timber Research and Development Association (TRADA), High Wycombe, UK: 4.203–210.

Hayashi, T., Sasaki, H. and Masuda, M. (1980) 'Fatigue properties of wood butt joints with metal plate connectors', *Forest Products Journal*, **30**(2): 49–54.

Hoffmeyer, P. (2003) 'Strength under long-term loading', In: Eds. S. Thelandersson and H.J. Larsen, Timber Engineering, John Wiley & Sons, Chichester, UK.

Karacabeyli, E. (1988) 'Duration of load research for lumber in North America', *Proceedings International Conference on Timber Engineering*, Washington State University, USA: 1.380–389.

Karacabeyli, E. and Soltis, L.A. (1991) 'State-of-the-art report on duration of load research for lumber in North America', *Proceedings International Timber Engineering Conference*, Timber Research and Development Association, High Wycombe, UK: 4.141–155.

Kellogg, R.M. (1960) 'Effect of repeated loading on tensile properties of wood', *Forest Products Journal*, **10**(11): 586–594.

Kohara, M. and Okuyama, T. (1992) 'Mechanical responses of wood to repeated loading V: Effect of duration time and number of repetitions on the time to failure in bending', *Journal of Japanese Wood Research Society*, **38**(8): 753–758.

Kommers, W.J. (1943a) 'Effect of 5,000 cycles of repeated bending stresses on five-ply Sitka spruce plywood', Report No: 1305, Forest Products Laboratory, Madison, WI, USA.

Kommers, W.J. (1943b) 'Effect of ten repetitions of stress on the bending and compressive strengths of Sitka spruce and Douglas fir', Report No: 1320, Forest Products Laboratory, Madison, WI, USA.

Kommers, W.J. (1943c) 'The fatigue behaviour of wood and plywood subjected to repeated and reversed bending stresses', Report No: 1327, Forest Products Laboratory, Madison, WI, USA.

Kuipers, J. (1986) 'Effect of age and/or load on timber strength', Paper 19-6-1, Volume I, *Proceedings of CIB-Working Commission 18: Timber Structures*, Meeting 19, International Council for Research and Innovation in Building and Construction, Rotterdam, The Netherlands.

Kyanka, G.H. (1980) 'Fatigue properties of wood and wood composites', *International Journal of Fracture*, **16**(6): 609–616.

Leicester, R.H. (1990) 'Failure in two-cycle fatigue', *Theoretical and Applied Fracture Mechanics*, **13**: 161–164.

Lewis, W.C. (1960) 'Design considerations for fatigue in timber structures', *ASCE Journal of the Structural Division*, **86**(ST5): 15–23.

Lewis, W.C. (1962) 'Fatigue resistance of quarter-scale bridge stringers in flexure and shear', Report No: 2236, US Forest Products Laboratory, Madison, WI, USA.

Madsen, B. (1992) Structural Behaviour of Timber, Timber Engineering Ltd., North Vancouver, BC, Canada.

McNatt, J.D. (1970) 'Design stresses for hardboard — Effect of rate, duration, and repeated loading', *Forest Products Journal*, **20**(1): 53–60.

McNatt, J.D. and Werren, F. (1976) 'Fatigue properties of three particleboards in tension and interlaminar shear', *Forest Products Journal*, **26**(5), 45–48.

Mohammad, M.A.H. and Smith, I. (1994) 'Stiffness of nailed OSB-to-lumber connections', *Forest Products Journal*, **44**(11/12): 37–44.

Mohammad, M.A.H. and Smith, I. (1996) 'Effects of multi-phase moisture conditioning on stiffness of nailed OSB-to-lumber connections', *Forest Products Journal*, **46**(4): 76–83.

Nielsen, L.F. (1996) 'Lifetime and residual strength of wood', Report: Series R, No. 6, Department of Structural Engineering and Materials, Technical University of Denmark, Lyngby, Denmark.

Okuyama, T., Itoh, A. and Marsoem, S.N. (1984) 'Mechanical responses of wood to repeated loading I: Tensile and compressive fatigue fractures', *Journal of Japanese Wood Research Society*, **30**(10): 791–798.

Puskar, A. and Golovin, S.A. (1985) Fatigue in Materials: Cumulative damage Processes, Publishing House of the Slovak Academy of Science, Bratislava.

Rose, G. (1965) 'The mechanical behaviour of pinewood under dynamic constant stress depending on kind and amount of load, moisture content, and temperature', *Holz als Roh-und Werkstoff*, **23**(7): 271–284.

Rosenthal, D. (1964) Introduction to Properties of Materials, D. Van Nostrand, Princeton, NJ.

Sekhar, A.C. and Shukla, N.K. (1979) 'Some studies on the influence of specific gravity on fatigue strength of Indian timbers', *Journal of the Indian Academy of Wood Science*, **10**(1): 1–5.

Sieminski, R. (1960) 'Fatigue strength of pinewood (Pinus sylvestris)', *Holz als Roh-und Werkstoff*, **18**(10): 369–377.

Smith, I. and Gong, M. (2002) 'Fatigue in spruce lumber under compression parallel to grain', *Proceedings of World Conference on Timber Engineering*, MARA University of Technology, Shah Alam, Malaysia, **3**: 61–68.

Smith, I., Daneff, G, Ni, C. and Chui, Y.H. (1998) 'Performance of bolted and nailed timber connections subjected to seismic loading', In Special Publication 7275, Forest Products Society, Madison, WI, USA: 6–17.

Tsai, K.T. and Ansell, M.P. (1990) 'The fatigue properties of wood in flexure', *Journal of Material Science*, **25**: 865–878.

Whale, L.R.J. (1988) 'Deformation characteristics of nailed or bolted timber joints subjected to irregular short or medium term lateral loading', PhD thesis, Polytechnic of the South Bank, London, UK.

Wood, L. (1951) 'Relation of strength of wood to duration of load', Report No. 1916, US Forest Products Laboratory, Madison, WI, USA.

Zhou, H. and Smith, I. (1991) 'Influences of drying treatments on bending properties of plantation — grown white spruce', *Forest Products Journal*, **41** (2): 8–14.

Appendix: Notation

AD = accumulated deformation

C_{max} = static compressive strength

HCF = high cycle fatigue

FL = strength ratio

i = specimen rank

m	= MC	= moisture content (%)
n		= number of load cycles
n_r		= number of replicates
N		= fatigue life (number of repetitive load cycles to failure)
LCF		= low cycle fatigue
LVL		= laminated veneer lumber
PSL		= parallel strand lumber
R-ratio		= stress ratio
R_{crit}		= critical value of R-ratio
S		= stress
t		= total loading time
T		= lifetime (time to failure under sustained load)
σ		= stress level
σ_a		= alternating stress
σ_e		= stress amplitude at R-ratio = 1.0
σ_m		= mean stress
σ_0		= threshold stress level
σ_u		= ultimate strength
τ		= duty ratio

7

Fracture Modelling in Wood

7.1 The Modelling Problem

As with nearly every scientific or engineering problem, the fundamental goal of fracture modelling in wood is to fashion the capability to make general predictions of material behaviour based on a limited number of tests. Descriptions of fracture mechanics in Chapters 4 and 5 focused on experiments with laboratory-sized specimens. While work in the laboratory provides useful insight into how the material responds under a variety of load states, this insight may not necessarily tell us how to predict the performance of large structural systems. It is the goal of all structural models to make such connections between small scale and large scale behaviour.

In a homogeneous material, such as most metals, fundamental material characteristics are essentially all contained at the scale of the crystalline structure. This means that at any scale above this basic level, the material may be considered a continuum, and only a few parameters are necessary to predict material performance. The exception is the case when the material contains cracks. However, basic fracture mechanics theory provides the tools to predict strength and toughness of cracked materials.

For wood this connection is extremely problematic because of the wide range of 'material length scales'. Material features that affect performance can be observed at scales ranging from centimetres (knots and checks) down to nanometres (microfibrils). A material model fine enough to capture effects of microfibril angle will quickly become too large to be able to make predictions at the scale of a structural element. By contrast, a model that can quantify the impact of visible defects will be too coarse to include the effects of features such as pits in the cell wall.

Consideration of Figure 7.1 can further emphasize this problem of length scale. To predict the performance of a large structure, one needs property information for the individual components. The property information at the component level, however, is dependent on distribution of knots, grain angles and deviations, earlywood-latewood transitions, cell wall properties, and microfibril geometry and crystallinity. These factors cover a length range approaching nine orders of magnitude! To date, no approach has been truly able to bridge all these length scales. The typical approach is to choose

Fracture and Fatigue in Wood   I. Smith, E. Landis and M. Gong
© 2003 John Wiley & Sons, Ltd   ISBN: 0-471-48708-2 (HB)

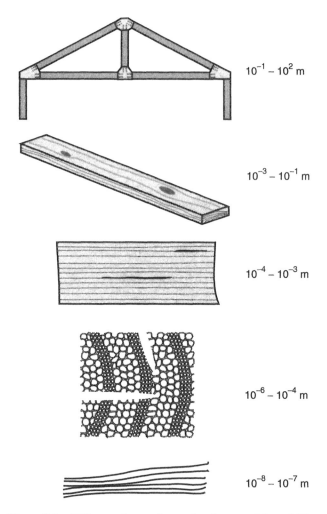

$10^{-1} - 10^2$ m

$10^{-3} - 10^{-1}$ m

$10^{-4} - 10^{-3}$ m

$10^{-6} - 10^{-4}$ m

$10^{-8} - 10^{-7}$ m

Figure 7.1   Different observation scales for material modelling.

a scale of interest, and focus on deterministic or stochastic ways to account for the effects at other scales.

This chapter focuses on examples of different modelling approaches spanning the different ranges of length scale. These approaches included Weibull models, where length scale is treated in a purely statistical manner, nonlinear fracture models that use a fracture mechanics basis, and a variety of computational approaches based on both continuum and non-continuum representations.

7.2   Statistical Fracture Models

Fracture mechanics, as presented in this text, is a deterministic theory. Through the use of a continuum framework, relationships between structure, load, crack size, stress intensity, and energy release rates can be derived, and applied to the prediction of strength and toughness. However, the relationship between a material's cracks and its strength can be described in purely statistical terms as well. Such models have been found particularly useful in predicting size effects observed in many materials, and have been applied in predictive models for wood strength by a number of researchers. Because of their popularity, a brief development of statistical failure theory is presented here. It should be kept in mind, however, that there are some compelling arguments that can be made that suggest the basis for applications to wood are questionable. Such arguments are discussed below.

The most popularly cited statistical fracture theory is credited to Weibull (1939), but roots of the theory can be traced to work by da Vinci, Galileo and Mariotte. Consistent with modern fracture mechanics, Weibull recognised that a critical flaw in the material dictates failure strength. As with 'weakest-link' theories, the greater the likelihood of a critical flaw existing in a material, the greater is the likelihood of failure at a particular stress. From this idea, it logically follows that the measured strength distribution of a material is determined by the probability that a critical flaw will be present. Weibull developed a model distribution that is appropriate for the extreme value statistics of material strength.

Mathematically, the theory can be illustrated as follows. Consider a bar made up of many Representative Volume Elements (RVEs), as shown in Figure 7.2. If it is assumed that each RVE has a cumulative strength distribution $p_0(\sigma)$, then the probability that the element will not fail at a given stress is equal to $1 - p_0(\sigma)$. According to the theorem of joint probabilities, the probability that the bar will not fail at the stress level can be written as the product of all the element probabilities:

$$1 - p_f = (1 - p_0)(1 - p_0)(1 - p_0)\ldots = (1 - p_0)^N \tag{7.1}$$

where p_f is the probability the bar will fail. Taking the natural logarithm of both sides yields:

$$\ln(1 - p_f) = N \ln(1 - p_0) \tag{7.2}$$

Recognising that p_0 must be small for small elements, the right hand side of Equation (7.2) may be approximated as:

$$\ln(1 - p_0) \approx -p_0 \tag{7.3}$$

Substituting this result into Equation (7.2) and solving for p_f yields:

$$p_f = 1 - e^{-N p_0(\sigma)} \tag{7.4}$$

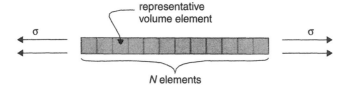

Figure 7.2   Material sample made up of N discrete elements.

Equation (7.4) can be rewritten for a specimen of arbitrary geometry by expressing it in terms of the volume of the specimen relative to the volume of the RVE. Let V = the volume of the specimen, and V_0 the volume of the RVE. Then $N = V/V_0$, and Equation (7.4) can be written as:

$$p_f = 1 - e^{-\frac{V}{V_0} p_0(\sigma)} \tag{7.5}$$

In this form it can clearly be seen that as the volume of the specimen, V, is increased, the probability of the structure failing also increases, an observation that can be traced back to the time of da Vinci.

Equation (7.5) leaves us with the critical problem of the probability of failure, p_0, in the RVE. Weibull (1939, 1951) suggested the following formula for p_0:

$$p_0(\sigma) = \left(\frac{\sigma - \sigma_0}{\omega} \right)^m \tag{7.6}$$

where σ_0, ω, and m are empirically derived constants. The σ_0, is the strength threshold, or minimum possible strength. The ω is a scale parameter and m is a shape parameter (often referred to as the Weibull modulus). Substituting Equation (7.6) into Equation (7.5) yields a convenient form of the Weibull distribution:

$$p_f = 1 - e^{-\frac{V}{V_0} \left(\frac{\sigma - \sigma_0}{\omega} \right)^m} \tag{7.7}$$

Example plots of p_0 and p_f are shown in Figure 7.3. It should be noted that equation (7.7) assumes that p_0 is uniform throughout the structure. For materials with a spatially varying localised probability of failure, Equation (7.7) can be generalised to:

$$p_f = 1 - e^{-\int_V p_0(\mathbf{x}) dV} \tag{7.8}$$

where $p_0(\mathbf{x})$ is a the probability of failure of a local element at location $\mathbf{x}$.

Practical applications of Weibull theory to wood failure are common due to the attractive way it appears to explain observed size effects in terms of local material

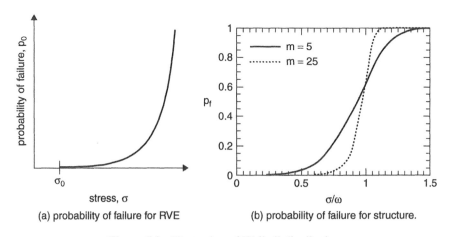

(a) probability of failure for RVE   (b) probability of failure for structure.

Figure 7.3   Illustration of Weibull distribution.

properties. Experimental evaluation of the three constants σ_0, ω and m provides the necessary information to make statistically significant predictions of specimen strength. Frequently, the minimum or threshold strength, σ_0, is set equal to zero, making Equation (7.7) a two parameter rather than a three parameter model.

In what has been perhaps the most comprehensive application to Weibull theory to wood, Barrett (1974) derived a model to assess the size effects in Douglas-fir loaded in tension perpendicular to the grain. For geometrically similar specimens of varying volume, the model predicted that the log of specimen strength, σ_{max} will decrease linearly with the log of specimen volume, V:

$$\log \sigma_{max} = a - \frac{1}{m} \log V \qquad (7.9)$$

where a is a constant based on local failure probabilities. (Note that 10^a represents the mean strength of a unit volume of material.) Size effects were examined in specimens of varying quality. The Weibull modulus, m, was found to be about 5, while the strength of a unit volume of material was found to be about 3.2 MPa (460 psi). The model provided a reasonably good fit with experimental data, and the results were used to suggest volume correction factors for design standards.

Most of the progress made over the last 40 years has been towards applications to a variety of loading conditions and materials, as well as refinements of the statistical basis, and implementation of the theory's predictions to design standards. Examples of the first case include Liu's (1982) evaluation of size effects in wood flexure specimens using a two-parameter model, and Marx and Evans' (1986, 1988) use of a three-parameter Weibull distribution (among others) to evaluate strength distributions of laminating stock for use in reliability-based design of glued-laminated-timber (glulam) beams. Refinements of the statistical significance of applications of the Weibull distribution was conducted by Evans *et al.* (1998) to establish goodness-of-fit measures for the many applications of the distribution. Barrett *et al.* (1995) used Weibull theory to establish length and width strength adjustment factors for design of tension, bending, and compression structural members.

For materials in general, the RVE corresponds to the smallest volume of material that can realistically be considered a continuum. For wood, identifying the RVE is a problematic concept, as the size of heterogeneities can approach that of the structural element! The problem is solved in practice by empirically selecting a size of RVE that leads to a good calibration of Equation (7.7) against test data that relates apparent material strength to specimen volume. Unfortunately, the practice leads to inconsistent estimates of model parameters, meaning that it is unreliable to extrapolate predictions to volumes outside the range of the test data. Equation (7.8) points to a potential solution to the problem, but as is immediately apparent there are great practical difficulties associated with calibrating $p_0(\mathbf{x})$. Also, even if an element is sensibly one dimensional (tie rod in Figure 7.2), as discussed in Chapters 2 and 3 wood has a structure that is three-dimensional. This means that generalised forms of Weibull type theory need to be multi-dimensional.

Despite the vast number of applications of Weibull weakest link theory to wood (numbers well beyond the few cases cited here), there are some issues that bring the fundamental applicability into question. Bazant (1999) presents several objections to the application of classical Weibull theory to quasi-brittle structures. Implicit in Weibull theory is the relationship between flaw size and rupture. The theory presumes that a given flaw size equates with a certain critical load, which is an appropriate assumption for purely brittle materials, but not for materials that exhibit numerous toughening mechanisms. There is no accounting for the stress redistribution that can take place at a crack when different toughening mechanisms are mobilised. In addition, there is no account for microstructural length scales in the Weibull equations presented above. A homogeneous (albeit cracked) continuum is assumed in the theory. However, as already noted, the structure of wood is such that material features covering a wide range of length scales affect fracture behaviour.

As discussed in Section 2.2, wood can contain quite extensive damage of various types and from various causes, but there is no basis on which to presume that it is homogenously dispersed in a truly statistic sense. As pointed out by Boatright and Garrett (1980), there is not necessarily a relationship between the critical flaw sizes calculated from tensile tests of clear specimens, and the real flaws in wood. They further showed that because fracture is typically confined to planes of weakness along the grain, strength cannot be truly random throughout the volume. And after all, that is something which is obvious if one knows anything about tree growth and physiology.

Despite above criticisms, it would be unfair to imply that Weibull theory has no place in one's modelling arsenal. Bazant suggested the theory is perhaps appropriate for large structures where the effects of toughening mechanisms are not as strongly felt. (He noted that for structures such as concrete dams, the theory is applicable.) Also, it is fair to say that recognition of size effects is important when using conventional strength-based failure criteria. For example, Clouston *et al.* (1998) implemented weakest-link theory with Tsai-Wu failure theory (see Chapter 3) to assess volume effects in Laminated Veneer Lumber. By integrating these two approaches, size effects can be introduced into a general stress state failure criterion that would not otherwise predict the effects.

Regarding practical application of Weibull and similar theories, capacities can be under predicted by a considerable extent when analysed components exhibit stress concentrations (e.g. drilled or notched members, mechanical connections). This is mainly for two reasons. First, predictions usually are made for one stress component at a time, and only for states of stress expected to exhibit approximately brittle-elastic characteristics (tension and shear). There is no allowance for softening of stress peaks as often occurs under complex multi-axial stress states that accompany stress concentrations and promote non-linearity. Second, it is usually assumed when applying equations such as Equation (7.8) that probability of failure is an aggregation of probabilities for all increments of volume that make up a component or entire structural system. This is unrealistic when there are stress concentrations that are loci for major fracturing and the location(s) of the failure plane(s) is for all practical intents constrained. Although it is often argued that contributions that individual lightly stressed volume increments make to p_f are small, their collective influence can be large. This source of discrepancy can be particularly acute if the two-parameter form of the Weibull theory is employed, which is often the case as calculations and calibration of model parameters are greatly simplified.

Another consideration when applying Weibull type theory is whether the material is sensibly a homogenous continuum. For solid wood this depends upon factors such as species and absolute size of a component. For wood based composites factors such as particle size are an issue. In practice structural features of the material need to be at least one order of magnitude smaller that dimensions of regions that develop steep stress gradients. Certain tropical hardwoods are relatively quite homogenous and this is not an issue. Unfortunately, the same is not true in other cases. For example, particles in Oriented Strand Board (OSB) and Laminated Strand Lumber (LSL) are often large compared with dimensions of high intensity stress regions that lie adjacent to glued or mechanical joint planes. As for any theory, there are strategies that mitigate problems associated with application of statistical fracture theories. These include, as implied above, use of models that recognise that materials have none zero threshold strength, recognition that all regions of a member are not equally liable to be the site of the fracture plane, and abstinence from applying the theory to coarsely 'grained' materials with stress concentrations.

7.3   Nonlinear Fracture Mechanics Modelling

Fracture mechanics theory provides a basis for relating the effects of cracks to the bulk strength and toughness of a material. As detailed in Chapter 4, crack size is related to bulk properties through application of stress intensity factors, strain energy release rates, and other tools of Linear Elastic Fracture Mechanics (LEFM). These tools can be used to make performance predictions for full-scale structures. However, as pointed out in Chapter 5, wood tends to exhibit toughening mechanisms that cause LEFM solutions to be inaccurate. Thus, there has been significant interest in developing nonlinear fracture models that can more accurately capture the effect of the toughening mechanisms, and better predict fracture strength and toughness.

7.3.1   Fictitious crack models

The most common nonlinear fracture mechanics modelling approach for wood has been the application of fictitious or cohesive crack models, specifically, variations of a model by Hillerborg *et al.* (1976) and Hillerborg (1991). As described in Chapter 4 (Section 4.4.3), the fictitious crack approach introduces a cohesive zone at the tip of the crack. In order to advance, the crack must not only overcome crack resistance of the material, but also the stresses produced by the cohesive zone. The fracture behaviour of the material can be characterised by the response of the material to a uniaxial tension force. As illustrated in Figure 7.4, the deformation in the specimen is made up of two components: the elastic strain, and the crack opening displacement, w. Up to peak load the elastic deformation dominates, but after the peak, the elastic strains disappear, and the deformation is due primarily to w. The function $\sigma(w)$ is considered to be a material property, and the fracture energy G_f (in this case the energy required to completely separate two faces of the material) is the area under the $\sigma(w)$ curve. Application of the fictitious crack model is typically through finite element analysis. Crack elements with predetermined softening properties are incorporated into the structural model.

Emphasis has been on mode I fracture, with Bostrom (1992) appearing to be the first person to apply fictitious crack concepts to wood. From deformation controlled tests on Scots Pine in direct tension and shear, he surmised that the fracture process zone in wood could be characterised as behaving in a similar way to concrete. Accordingly, he determined post peak softening parameters and fracture energy. Size effects, as well

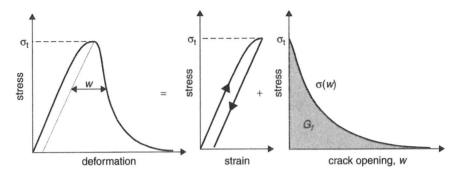

Figure 7.4   Basis for the fictitious crack model.

as experimental problems involved in making these measurements were recognised. Because of the experimental difficulties in measuring post-peak softening in direct tension, Stanzl-Tschegg *et al.* (1995) developed a wedge-splitting test protocol. An illustration of their specimen is shown in Figure 7.5. A notched specimen is split apart by a wedge that pushes against rollered steel platens. Like the double cantilever beam specimen under displacement control discussed in Chapter 4, this specimen configuration guarantees a stable crack growth because strain energy release rate, G, decreases as the crack extends. Thus, the wedge must be continually pushed deeper into the specimen to further extend the crack. Load-displacement curves similar to the one shown in Figure 7.4 are produced.

Because the geometry adopted by Stanzl-Tschegg *et al.* does not produce a pure tensile state assumed in the Hillerborg model, they developed a finite element model

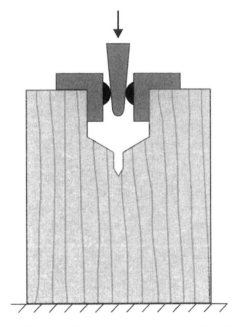

Figure 7.5   Illustration of wedge-splitting specimen used by Stanzl-Tschegg *et al.* (1995).

of the crack tip to separate tensile from bending components in their measurements. In their implementation of the Hillerborg model, a bilinear softening diagram was assumed, as is illustrated in Figure 7.6. They further assumed, based on published studies of other strain-softening materials, that the fracture energy is composed of contributions from both microcracking (ahead of the crack tip) and crack bridging (behind the crack tip). As shown in the figure, areas defined by two stresses and two displacement values can determine the two fracture energy terms. The bilinear diagrams were determined through an iterative comparison between the finite element model predictions and the experimentally measured load-deformation responses. Examples of such curves are shown in Figure 7.7. It is interesting to note that the 'bridging' component has a much larger relative contribution to total fracture energy in the RL direction than in the TL. This reflects the presence of ray cells normal to the crack plane for the RL direction.

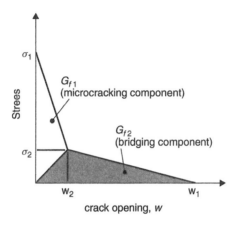

Figure 7.6 Bilinear $\sigma(w)$ diagram with microcracking and bridging components of fracture energy illustrated.

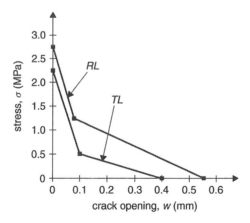

Figure 7.7 Example $\sigma(w)$ diagrams for RL and TL fracture tests.

7.3.2 Bridging model

Because of uncertainties of the exact nature of fracture, Vasic and Smith (2002) made a comprehensive study of mode I fracture processes in the RL and TL directions using Scanning Electron Microscope (SEM) observations combined with finite element simulations. Small wedge splitting specimens of Eastern Canadian spruce were loaded inside an SEM chamber, and accompanying changes in microstructure were recorded on videotape. Some of the SEM observations are as follows. First, the crack tip is difficult to distinguish amidst the complex structure of the material. Secondly, both damage ahead of the crack tip as well as fibre bridging behind the crack tip were observed. Thirdly, bridging appeared to be the mechanism most responsible for the nonlinear fracture response. Quantitatively, they were able to reconstruct crack tip profiles from the video recordings. Based on comparisons with LEFM solutions for crack tip displacements they were able to establish bridging stresses, an example of which is shown in Figure 7.8.

Based on this work, they established a crack bridging model that is similar in practice to the fictitious crack model, however a sharp crack tip with a stress singularity can exist in the bridging model. In addition, the bridging zone is considered a real material feature, rather than the fictitious zone assumed by Hillerborg. An example of the model parameters is shown in Figure 7.9. The parameters include the length of the bridging zone, the crack opening displacement at which the zone ceases to act, and the stress intensity factor at the crack tip. Mathematically the bridging model is:

$$K_{\text{net}} = K_o + K_b \tag{7.10}$$

where K_{net} is the total stress derived toughness (with bridging), K_o is the intrinsic toughness at the crack tip (for a sharp crack without bridging), and K_b is the bridging toughness. This crack-bridging model was implemented in a finite element simulation of a wedge loaded spruce specimen. The most interesting finding is that the effect of the bridging zone reached a maximum at a bridging zone length, L_{br}, of about 4 mm. This can be compared with the typical length of tracheid cells in Eastern Canadian spruce. Thus there is a physical relationship between the size of the bridging zone, and an

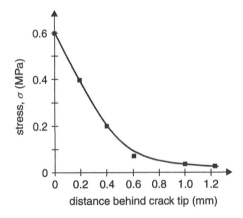

Figure 7.8   Calculated bridging stresses acting behind the crack tip (RL direction).

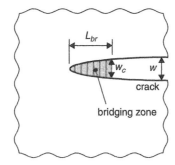

Figure 7.9   Schematic illustration of a bridged crack model.

intrinsic material feature. The explanation is that dispersed micro-cracking associated with inherent damage (see Section 2.2) does not exceed tracheid length because such cracking is arrested at where cell caps interlock.

7.3.3   Applications

Applications of fictitious crack or bridging models to practical engineering problems have been somewhat scarce. In one recent example, Serrano *et al.* (2001) applied fictitious crack concepts to the problem of modelling finger joints in lamination stock. They essentially treated the joints as nonlinear springs with a stress-elongation diagram as illustrated in Figure 7.10. A trilinear relationship was defined in order to capture the

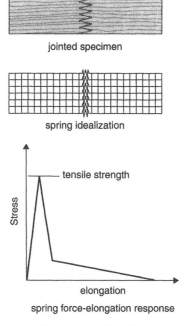

Figure 7.10   Illustration of model features used by Serrano *et al.* (2001) for finger joints in lamination stock.

essential features of experimental data, namely two distinct levels of strain softening. Material outside the region of the joint was treated as a linear elastic, orthotropic continuum, making finite element implementation fairly straightforward.

In their analysis, they varied the characteristics defining the stress-elongation diagrams, and used Monte Carlo methods to simulate the influence of different parameters on model predictions. In addition to their model predictions agreeing fairly well with experimental data, the model predicted size effects and beam depth effects.

While it has been fairly well established that nonlinear effects must be accounted for in fracture analysis of wood, the subject is relatively unstudied. It is likely that strictly deterministic material models will never adequately describe the behaviours readily observed in wood fracture. However, it should be equally clear that purely stochastic models will not capture the complex mechanisms contributing to the observed behaviour. It seems that what is called for is a hybrid method that can capture the complexity of different fracture mechanisms, but can equally capture the wide range of property variation seen both among and within species groups, and within individual structural components.

7.4   Other Modelling Paradigms

7.4.1   Finite element models

In the previous section it is stated that nonlinear fracture models are necessary to capture effects of material toughening mechanisms. However, it is worth noting there have been some fairly innovative modelling approaches that while employing LEFM fracture criteria, are able to predict some important fracture characteristics of wood. Zandbergs and Smith (1988) and Cramer and Fohrell (1990) employed continuum-based finite element methods to simulate performance of wood members subjected to different stress states. The innovation in their work is the way they treated the grain orientation and the presence of knots. In both cases, a 'flow-grain analogy', introduced by Phillips *et al.* (1981), mimiced the geometry of wood grain around knots.

As illustrated in Figure 7.11 the 2D finite element mesh generated by STARWX (Zandbergs and Smith, 1988) is geometrically similar to the grain observed in typical dimension lumber. Longitudinal 'flow lines' are established using fluid mechanics equations for laminar flow around an elliptical cylinder. Grain angle relative to the direction of tensile stress is arbitrarily set and a mixed mode fracture criterion is used (Wu, 1967) to predict local fracture. The model is reasonably accurate at predicting test results, including region of crack initiation, peak load and progressive failure patterns. The analysis, however, does not predict post-peak load-deformation behaviour well.

In Cramer and Fohrell's work, the goal was to improve lumber grading procedures through a more rational representation of the material. Model parameters were established by small-scale specimen testing, and grain-angle maps were made of lumber specimens to determine specimen-specific material features. Boards of both Douglas fir and southern pine were mapped, simulated and tested in tension parallel to grain. The two species were subjected to different end conditions during testing, the Douglas fir being constrained at the end, while the southern pine was allowed to rotate at the ends. As shown in Figure 7.12a, the model made very good predictions of strength. In addition, failure patterns, and post-peak load-deformation behaviour were reasonably

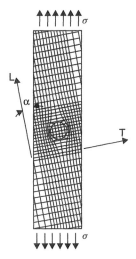

Figure 7.11 Finite element mesh used by Zandbergs and Smith (1988) to simulate effects of knots and arbitrary grain angles.

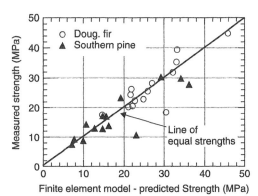

(a) Comparison of morphology-based finite element model predictions with experimental measurements.

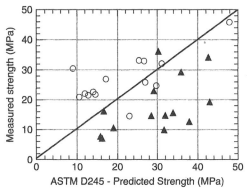

(b) Comparison of ASTM D245 predictions with experimental measurements.

Figure 7.12 Predictions and experimental measurements from Cramer and Fohrell (1990).

well simulated. Of note is the comparison between the measured strengths, and the strengths predicted using the empirical methods of ASTM D245 (ASTM 1997). As seen in Figure 7.12b, the results are not quite as good. The finite element model based on fundamental material characteristics clearly serves as a better predictor than a purely empirical approach. No doubt there is a trade-off on ease of use, however.

These studies represent an important contribution in that the models used are formulated from basic principles. Material response to load is predicted without reliance on empirical correction factors. Predicted material variability is due to measurable structural features. The model, in theory, could thus be applied to the strength prediction of in-service structures if appropriate grain geometry and material properties could be properly established.

7.4.2   Morphology-based models

As the name implies, a morphology-based model is one where the parameters of the computational model are drawn directly from the physical structure of the material. Rather than treating the material as a statistically homogenized continuum, it is represented as a collection of discrete elements each with assigned properties based on the different microstructural features represented.

The previously discussed finite element simulations represent a first step in a morphological model basis. The structure of the material at a visible length scale is represented by the geometry of the finite element mesh. Material properties are taken from small scale clear test specimens, and factors such as density can be simulated by appropriate spacing of grain lines.

Gibson and Ashby (1988) carried the idea a step further by working at an even finer length scale. They modelled wood structure as a parallel array of closely packed hexagonal prisms, stiffened by rays, end caps, and transverse membranes. The model gives characteristic material properties based on the density of the wood and the direction of loading. Elastic properties, strength and fracture toughness are analysed based on the assumption that the composition and lay-up of wood microstructure in different woods is similar, and that the differences between species can be traced to their cellular structure. This is an entirely rational approach that can be implemented based on current knowledge. Unfortunately the approach has not gained wide acceptance because it tends to be very computationally expensive. Simplifying assumptions related to cell structure, such as taking the cross-section to be regular and of average dimension do lead to substantial computational economies and can make numerical techniques unnecessary (see Chapter 3). However, as discussed in Section 3.3, simplifications that ignore irregular structure cannot be expected to yield acceptable predictions of failure behaviour.

7.4.3   Lattice models

Lattice models can be considered as an alternative to the two previous modelling approaches. With a lattice material model, the traditional continuum representation of the material is abandoned, and the material is represented by an array (or lattice) of interconnected discrete bar elements. Properties of the elements can correspond to physical microstructural features. Lattice models have been common for a variety of

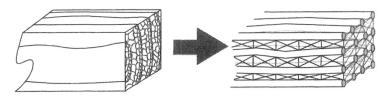

Figure 7.13   Illustration of a lattice model for wood.

heterogeneous materials (Herrmann and Roux, 1990), because of the simple way that disorder can be easily introduced into fracture and damage problems.

An example of a lattice application to wood is shown in Figure 7.13, where the bundles of tracheids or fibres in the xylem are represented by tubular beam elements connected by a lattice of springs. Through this type of representation, material defects such as knots, cracks and grain angle deviations can be incorporated directly into the model in a manner similar to the flow-grain analogy previously described. In this case, additional disorder is added through assignment of element properties based on a given statistical distribution.

For a material such as wood, where heterogeneities cover such a wide range of length scales, a lattice model offers several advantages over conventional continuum representations. First, the model can represent the material in a way that has a physical basis (Figure 7.13). Material elements are arranged in a lattice that mimics the structure of the real material. Local variations in fiber orientation, shakes and checks, and other heterogeneities or discontinuities can be represented directly in the lattice. Secondly, the changes in microstructural features that arise from damage inducing mechanisms can be handled explicitly through broken elements. Thirdly, and most significantly, the modelling approach establishes a computational link between the bulk behaviour of the material to the micro- and meso-level features that cause that behaviour. Damage, defects, or other microstructural features can be quantifiably linked to changes in material stiffness and strength. The model should capture damage patterns that result from arbitrary loading states along with the bulk load-deformation response.

The goal of this modelling approach is to be able to simultaneously mimic both material damage phenomena and bulk material properties. If the model is truly able to capture the physical processes involved, then results of such hard-to-model phenomena as damage-induced strain softening, interacting cracks, and volume effects are predicted explicitly. Furthermore, this approach lends itself to directly integrating experimental measurements. Crack distributions predicted by the model can be measured experimentally.

In representing the material in this way, a common paradox in modelling heterogeneous materials is avoided. Traditionally, a heterogeneous medium is first homogenised to average properties over a representative volume, and then it is discretised to permit a finite element solution. In the approach described here, the heterogeneous medium is represented as a heterogeneous medium eliminating the errors associated with both homogenisation and discretisation.

Landis *et al.* (2002) and Davids *et al.* (2003) examined aspects of 2D lattice models as a way to link microstructural characteristics to bulk mechanical properties. In their work, experimental results of Vasic (2000) were simulated using the 2D lattice shown in Figure 7.14. Specifically, tension perpendicular to the grain (RL direction) was

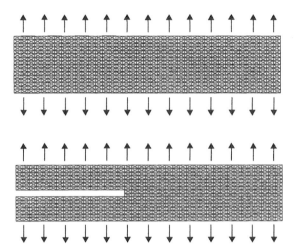

Figure 7.14   2D lattice used to simulate tension perpendicular to the grain in un-notched (top) and notched (bottom) specimens.

considered in both notched and un-notched specimens. In these simulations, cylindrical flexure elements were used to simulate microstructure in the longitudinal direction. These elements were connected by an array of springs oriented in the (R) direction. In addition, to simulate Poisson effects, diagonal springs were inserted between the longitudinal elements. Strength and stiffness properties of all elements were randomised to simulate material variability at a scale finer than the scale of the mesh. At this point, the randomisation is set to match experimental measurements of bulk properties, but it could be tied to either very small-scale property measurements, or some other rational assumptions of fundamental characteristics.

Simulation of load-deformation response was conducted by applying a known displacement at the upper edge of the lattice, while holding the bottom edge fixed. Element forces were calculated using a traditional stiffness method of structural analysis. Once equilibrium was established, the forces in all elements were checked against their established strengths. If the force in the element exceeded the strength, the element was removed (stiffness set equal to zero), and equilibrium was recalculated. If no additional elements break as a result of the newly established equilibrium, the total force required to hold the prescribed displacement is calculated. This process is repeated at successively higher displacements until the specimen can no longer sustain a load.

Representative load-deformation curves are shown in Figure 7.15 while the corresponding simulated damage patterns are shown in Figure 7.16. Clearly, the method is able to capture non-linearity in the pre-peak region, as well as strain softening in the post-peak region. The damage patterns show microcracking, crack branching and bridging in addition to straight segments along the grain. As would be expected, damage in notched specimens is constrained to occur mainly in a plane lying ahead of the notch. Although damage in specimens without a notch tends to be more dispersed, there is a dominant fracture plane associated with the weakest tangentially oriented layer which in physical specimens is the weakest layer through the thickness of growth rings.

In subsequent work with lattices by Parrod (2002), additional stress states such as tension and shear parallel to grain, and bending were considered. An example

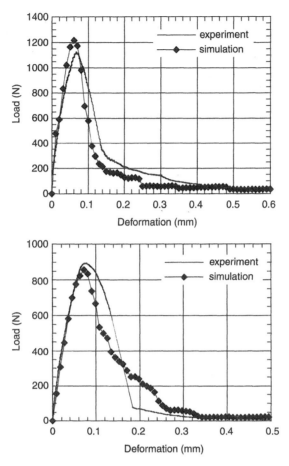

Figure 7.15 Typical load-deformation curves for experiments and lattice simulations of un-notched (top) and notched (bottom) tension perpendicular to grain specimens.

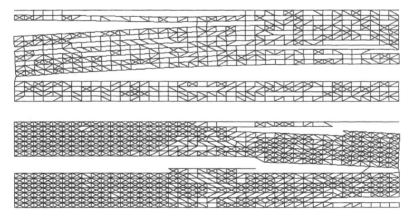

Figure 7.16 Damaged lattices for un-notched (top) and notched (bottom) tension perpendicular to grain specimens.

Figure 7.17   Simulated damage in small flexure specimen.

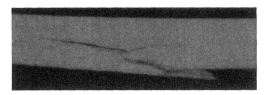

Figure 7.18   Picture of failure in a small flexure specimen, qualitatively similar to simulated specimen of Figure 7.17.

simulation of a bending specimen is shown in Figure 7.17. The damage pattern shows the traditionally observed fracture: tensile fracture at bottom fiber followed by peeling tension and shear along the grain. The simulated patter can be compared with an image of a small damaged flexure specimen shown in Figure 7.18.

The obvious drawback to the lattice simulation approach, as with Gibson and Ashby type morphological models, is the extremely large computational requirement for simulation of even a simple wood member. Likely future directions will couple these types of models with conventional orthotropic continuum-based finite element models. The reasonable approach is to implement the lattice model only in locations where damage occurs, as conventional finite element models are entirely appropriate for linear elastic behaviour.

However, it must be emphasized that substitution of a continuous medium in place of the heterogeneous material must be done with care. As has been illustrated in this and other chapters, the presence of 'anomalies' such as vessels, rays and other commonly found features can have a profound impact on bulk material behaviour. If these features are not considered in analysis, the result is the extremely wide statistical distributions of material properties of which people have (perhaps all too easily) grown accustom.

7.4.4 Damage models

Basic concepts of continuum damage mechanics were introduced in Chapter 4 (Section 4.6) as a phenomenological way of handling progressive cracking and damage that occurs in many materials. A damage variable is introduced that accounts for the degradation of material stiffness that occurs when an array of small cracks develops in the material.

Applications of continuum damage mechanics to wood have been sparse due probably to the difficulty in covering the wide range of length scales in a continuum framework. However an example of the technique is some work done by Faessel *et al.* (1999). Their work was based primarily on experimental work of Navi *et al.* (1995) who tested small uniaxial tension specimens loaded parallel to the grain. The wood is represented as a fibre-reinforced composite with wood cells embedded in an isotropic matrix. A fourth-order damage tensor relates the specimen stress to strain, and contains information on a number of different damage sources: fracture of individual fibres, which affect stiffness in the longitudinal direction, and delamination between fibres, which affects stiffness in the radial direction.

Details of their methods and results are not presented here, but their work is noteworthy in that it has the capacity to couple many different phenomena affecting fracture behaviour of wood. Namely visco-elastic, mechano-sorptive, damage, and fracture effects can (in theory) be modelled in a single damage formulation. In many respects damage models can be thought of as continuum versions of lattice models.

7.5 Which Model to Use?

The only thing that should be clear by now is that a wide variety of modelling approaches have been applied to the problem of fracture in wood, but no single universal method has emerged from the group. While research continues to be active on a number of fronts, for now selection of an appropriate model for a particular application is still largely driven by the scales of observation introduced at the beginning of the chapter.

Let it be assumed that the largest (and most common) observation scale is that of a structural element. This is based on the fact that nearly all modern structural analysis techniques are based on assembly of structural stiffness or flexibility matrices established from individual member properties. At the scale of a structural member it seems that Weibull theory, despite its drawbacks, is the most popular. This is likely due to its simplicity, and due to the fact that it does not require knowledge of specific failure modes. Caution is necessary because, despite there being a theoretical basis, applications of Weibull theory are entirely empirical owing to the need to calibrate parameters based on 'best fit of model' to data procedures. Therefore, there is no basis for presuming that predictions of how volume influences apparent strength of wood components are accurate beyond the range of calibration data. Truth of this is self-evident, as despite model predictions, capacities of very large components or systems do not in practice tend toward zero.

If there is sufficient knowledge of failure modes, it is argued here that better predictions of performance can be made through application of nonlinear fracture models. Incorporating fictitious crack or crack bridging models into finite element models of components can lead to more accurate predictions of peak loads, deformations and energy dissipation. Having said this, there is need to decide when to, and when not to, use nonlinear fracture models rather than those based on LEFM concepts. LEFM analysis of wood works well in situations where the stress intensity region adjacent to a notch, or another type of stress raising feature, is small in comparison with overall dimensions of the member. Under such circumstances strain fields around crack tips are sensibly the same as those in material tests that characterise fracture toughness. It does not matter whether wood's actual fracture behaviour is truly linear-elastic. What does matter is that the material behaviour is treated in a consistent manner when material property test data and structural components are analysed. LEFM based models are inadequate in any instance where crack tips and associated high stress regions cannot sensibly be thought of as remote from boundary conditions. A suggested rule of thumb for judging this is that LEFM is valid when the inelastic process zone is confined to a sphere of radius $r < 0.02a$ at a crack tip, where a is the crack length or any dimension of the cracked body (Smith and Vasic, 2003).

If detailed analysis is desired, but the failure mode is uncertain, it is probably necessary to apply morphology-based models that explicitly incorporate material structure into the model framework. At this point in time, these need to be employed at a relatively local level, but as with all computational methods, improvements in hardware speed and refinements in algorithm development will continue to broaden the range of applicability.

Some example problems are given in Chapter 9 to illustrate what the authors regard as appropriate applications of fracture models to wood.

7.6   References

ASTM (American Society for Testing and Materials) (1997) 'Standard practice for establishing structural grades and allowable properties for structural lumber', *Designation: D245-93 Book of ASTM Standards*, Vol. 03.01, ASTM, Philadelphia, PA, USA.

Barrett, J.D. (1974) 'Effect of size on tension perpendicular-to-grain strength of douglas-fir', *Wood and Fiber*, **6**: 126–143.

Barrett, J.D., Lam, F. and Lau, W. (1995) 'Size effects in visually graded softwood structural lumber', *Journal of Materials In Civil Engineering*, **7**: 19–30.

Bazant, Z.P. (1999) 'Size effect on structural strength: A review', *Archive of Applied Mechanics*, **69**: 703–725.

Boatright, S.W.J. and Garrett, G.G. (1980) 'On the statistical approach to fracture toughness variations with specimen size in wood', *Engineering Fracture Mechanics*, **13**(1): 107–110.

Bostrom, L. (1992) *Method for determination of the softening behavior of wood and applicability of a non-linear fracture mechanics model*, Report TVM-1012, Lund University, Lund, Sweden.

Clouston, P., Lam, F. and Barrett, J.D. (1998) 'Incorporating size effects in the Tsai-Wu strength theory for douglas-fir laminated veneer', *Wood Science and Technology*, **32**: 215–226.

Cramer, S.M. and Fohrell, W.B. (1990) 'Method for simulating tension performance of lumber members', *Journal of Structural Engineering*, **116**(10): 2729–2747.

Davids, W.G., Landis, E.N. and Vasic, S. (2003) 'Lattice models for the prediction of load-induced damage in wood', *Wood and Fiber Science*, **35**(1): 120–134.

Evans, J.W., Johnson, R.A. and Green, D.W. (1998) *Two- and three-parameter Weibull goodness-of-fit tests*, USDA Forest Service Forest Products Laboratory, Research Paper FPL-RP-493, Madison, WI, USA.

Faessel, C., Navi, P. and Jirasek, M. (1999) '2-d anisotropic damage model for wood in tension', *Proceedings of Damage in Wood*, COST Action E8, 27-28 May, Bordeaux, France.

Gibson, L.J. and Ashby, M.F. (1988) Cellular Solids — Structure and Properties, Pergamon Press, Oxford, UK.

Herrmann, H.J. and Roux, S. (1990) 'Statistical Models for the Fracture of Disordered Media, North-Holland, Amsterdam, The Netherlands.

Hillerborg, A. (1991) 'Application of the ficticious crack model to different types of materials', *International Journal of Fracture*, **51**: 95–102.

Hillerborg, A., Modeer, M. and Petersson, P.E. (1976) 'Analysis of crack formation and crack growth in concrete by means of fracture mechanics and finite elements', *Cement and Concrete Research*, **6**: 773–782.

Landis, E.N., Vasic, S., Davids, W.G. and Parrod, P. (2002) 'Coupled experiments and simulations of microstructural damage in wood', *Experimental Mechanics*, **42**(4): 389–394.

Liu, J.Y. (1982) 'A Weibull analysis of wood member bending strength', *Journal of Mechanical Design, Transactions of the ASME*, **104**: 572–577.

Marx, C.M. and Evans, J.W. (1986) 'Tensile strength of AITC 302-24 grade tension lamina-tions', *Forest Products Journal*, **36**: 13–19.

Marx, C.M. and Evans, J.W. (1988) 'Tensile strength of laminating grades of lumber', *Forest Products Journal*, **38**: 6–14.

Navi, P., Rastogi, P.K., Gresse, V. and Tolou, A. (1995) 'Micromechanics of wood subjected to axial tension', *Wood Science and Technology*, **29**: 411–429.

Parrod, P. (2002) 'Application of lattice models to fibrous materials', M.S. Thesis, University of Maine, Orono, Maine, USA.

Phillips, G.E., Bodig, J. and Goodman, J.R. (1981) 'Flow-grain analogy', *Wood Fiber Science*, **14**(2): 55–64.

Serrano, E., Gustafsson, P.J. and Larsen, H.J. (2001) 'Modeling of finger-joint failure in glued-laminated timber beams', *Journal of Structural Engineering*, **127**: 914–921.

Smith, I. and Vasic, S. (2003) 'Fracture behaviour of softwood', *Mechanics of Materials* (in press).

Stanzl-Tschegg, S.E., Tan, D.M. and Tschegg, E.K. (1995) 'New splitting method for wood fracture characterization', *Wood Science and Technology*, **29**: 31–50.

Vasic, S. (2000) 'Applications of fracture mechanics to wood', PhD Thesis, University of New Brunswick, Fredericton, NB, Canada.

Vasic, S. and Smith, I. (2002) 'Bridging crack model for fracture of spruce', *Engineering Fracture Mechanics*, **69**: 745–760.

Weibull, W. (1939) 'A statistical theory of the strength of materials', *Proc. Royal Swedish Academy of Eng. Sci.*, **151**: 1–45.

Weibull, W. (1951) 'A statistical distribution function of wide applicability', *Journal of Applied Mechanics*, **18**: 293–297.

Wu, E.M. (1967) 'Application of fracture mechanics to anisotropic plates', *Journal of Applied Mechanics*, **34**(4): 967–974.

Zandbergs, J.G. and Smith, F.W. (1988) 'Finite element fracture prediction for wood with knots and cross grain', *Wood and Fiber Science*, **20**(1): 97–106.

Appendix: Notation

a = empirical constant

G = energy release rate

G_f = fracture energy

K = stress intensity factor

K_o = intrinsic toughness at the crack tip

K_b = the bridging toughness

K_{net} = total stress derived toughness (including bridging)

L = longitudinal direction

LEFM = linear elastic fracture mechanics

L_{br} = length of bridging zone

m = shape parameter—an empirical constant

p_f = probability of failure

$p_0(\sigma)$ = cumulative strength distribution for RVE

R = radial direction

RVE = representative volume element

T = tangential direction

V = Volume
V_0 = Volume of RVE
w = crack opening
w_c = crack opening at end of the bridging zone
ω = empirical constant
σ = stress
σ_0 = strength threshold
σ_t = tensile strength
$\sigma(w)$ = stress-crack opening function

8

Fatigue Modelling in Wood

8.1 General Considerations

All fatigue models for wood presume damage cannot be recovered once initiated, and that no damage exists prior to application of external influences that cause internal stress. The first presumption is unlikely to lead to any serious discrepancy between theory and fact, but the second is undoubtedly dubious. As discussed in Chapter 2, solid wood contains pre-damage (often called inherent damage) that develops in the living tree, during felling, during processing and initial drying. Inherent damage is largely responsible for the lack of coherence in fatigue data both within and between research studies (see Chapter 6). For some wood with glue composites, it can be expected that inherent damage is of much reduced extent, and modelling assumptions are more likely to be reasonable.

In this chapter, models are classified as empirical, phenomenological or mechanics based. Each is an extrapolation tool, with none having a rigorous theoretical basis. Like rheological models, fatigue models are always calibrated against experimental data prior to use, rather than having as input fundamental physical and mechanical properties. Predictions from fatigue models are aids to engineering judgement, and they should always be utilised with that in mind and awareness of their limitations.

8.2 Empirical Models

As discussed in Section 6.2.2, cyclic and static fatigue effects on damage are inter-active rather than cumulative. Factors such as the number of load cycles, cumulative residence period at the peak load and form of the loading function (load sequencing, waveform and loading frequency) all influence life expectancy. The Palgrem–Miner rule (Palgrem, 1924; Miner, 1945) is the oldest and simplest model that recognises this. The concept embodied is that damage due to various events or processes is linearly additive. For pure fatigue this leads to the relations that at failure:

$$\sum_{i=1}^{I} \left(\frac{n_i}{N_i} \right) = 1.0 \tag{8.1}$$

Fracture and Fatigue in Wood I. Smith, E. Landis and M. Gong
© 2003 John Wiley & Sons, Ltd   ISBN: 0-471-48708-2 (HB)

where N_i is fatigue life at peak stress level σ_i, and n_i is the number of load cycles at σ_i. Equation (8.1) has been applied extensively to metals and reinforced concrete subjected to non-reversed loading. For reversed loading of materials having different strength when loaded in positive and negative senses (e.g. tension and compression), it is necessary to modified the rule (Chung *et al.*, 1989):

$$\sum_{i=1}^{I} \left(\alpha_i^+ \frac{n_i^+}{N_i^+} + \alpha_i^- \frac{n_i^-}{N_i^-} \right) = 1.0 \tag{8.2}$$

where $+$ and $-$ are indicators of the loading sense, and α_i^+ and α_i^- are calibration constants. The Palgrem–Miner concept is also applied to account for the combined influences of static and cyclic fatigue:

$$\sum_{i=1}^{I} \left(\frac{t_i}{T_i} + \frac{n_i}{N_i} \right) = 1.0 \tag{8.3}$$

where t_i is total loading time at stress level σ_i, and T_i is the pure creep (static fatigue) lifetime at σ_i. The residual fractional lifetime is assumed not to depend upon the mix of cyclic or sustained loading that has already consumed fractional lifetime. By implication, any Palgrem–Miner type model presumes that damage mechanisms are the same and propagate consistently under cyclic and sustained load. Figure 8.1 illustrates application of Equation (8.1) in the simple case where fatigue is due to repetitive load cycles ($I = 1$). Combination of the concepts underpinning Equations (8.2) and (8.3) yields:

$$\sum_{i=1}^{I} \left(\alpha_i^+ \frac{n_i^+}{N_i^+} + \alpha_i^- \frac{n_i^-}{N_i^-} + \beta_i^+ \frac{t_i^+}{T_i^+} + \beta_i^- \frac{t_i^-}{T_i^-} \right) = 1.0 \tag{8.4}$$

where α and β values are calibrated constants. There is no evidence that an equation like this has ever been applied in the context of wood, but logic suggests it could be appropriate.

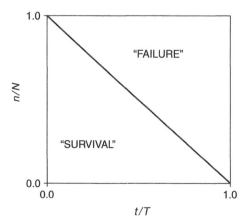

Figure 8.1 Palgrem–Miner rule applied to combined influences of static and cyclic fatigue under repetitive load cycles.

The Palgrem–Miner rule can lead to either excessively conservative or non-conservative predictions of fatigue life of fibre reinforced plastic laminates and joints in such materials (Sarkani, 1986; Sarkani *et al.*, 1999). For clear wood that fails in compression parallel to grain, a modified (non-linear) rule has been found most suitable in cases where damage results from repetitive load cycles (Kohara and Okuyama, 1992; Gong, 2000):

$$\left(\frac{t}{T}\right)^a + \left(\frac{n}{N}\right)^b = 1.0 \tag{8.5}$$

where a and b are calibrated constants. Beyond the context of repetitive load cycles, it would soon become impractical to apply a model of this type, unless both a and b are simple functions of σ.

As discussed extensively in Chapter 6, wood has memory of its stress history, and therefore the rate of damage accumulation under cyclic stress depends upon the load sequencing under variable amplitude stress cycles. This presents an insurmountable obstacle to any analyst attempting to apply simple rules, i.e. those that are insensitive to load sequence. An approach that might be able to mimic memory effects is the Sequence Dependent (SD) damage model (Sarkani and Lutes, 1988):

$$D_j = K^{-1}\sigma_j^m \left(\frac{\sigma_j}{\sigma_{j-1}}\right)^{\nu_1} \left(\frac{\sigma_j}{\sigma_{j-2}}\right)^{\nu_2} \cdots \left(\frac{\sigma_j}{\sigma_{j-i}}\right)^{\nu_i} \cdots \left(\frac{\sigma_j}{\sigma_{j-I}}\right)^{\nu_I} \tag{8.6}$$

where D_j is damage due to the jth cycle, $K^{-1}\sigma_j^m$ is a function defining the relationship between stress level σ and N (e.g. $= a\sigma^b$), σ_j is the stress level in jth cycle, and ν_i values are calibration constants. The SD model has been applied to welded joints in metals and is predicated on the assumption that damage in the jth cycle depends upon the stress range in the previous I cycles. Conceptually such an approach could apply to wood, including cases where the effect of creep needs to be included. Generalisations capable of handling effects of arbitrary load functions would have to be hereditary integral formulations. Although mathematically elegant, as was clarified by Whale (1988) in the context of rheological models, calibration of hereditary integral models is enormously complicated and the necessary extent of calibration data usually prohibitive.

State-variable models have been used to predict static fatigue in wood members. The simplest of these was proposed by Hendrickson *et al.* (1987) and calibrated against sustained constant flexural load data for clear Douglas-fir (Wood 1951). The model assumes the rate of consumption of the state variable to be:

$$\frac{dD}{dt} = A(\sigma - \sigma_o)^B \quad \text{if } \sigma > \sigma_o$$

$$\frac{dD}{dt} = 0 \qquad\qquad \text{if } \sigma \leq \sigma_o \tag{8.7}$$

where D is a damage index ($D = 0$ in undamaged sate, $D = 1$ at failure), σ is the applied stress level, σ_o is a threshold stress level below which no damage accumulates, and A and B are calibration constants. Barrett and Foschi (1978a) proposed a similar model for clear wood and structural lumber. They claim to have deduced their formulation from a materials science perspective, reasoning failure results from initiation, coalescence and propagation of microscopic and macroscopic voids or cracks, irrespective of the nature and duration of load. Time to failure is assumed to depend ultimately upon microstructure (this is only partly true if there is any gross pre-damage). Choice of an empirical rather than physical process model was justified based on the notion that static fatigue processes in wood products depend upon too complex an interplay of factors for other approaches to handle. Acceptability of such thinking depends on whether in practice wood is intelligent enough to smear influences that said complexities have on its behaviour. The premise might be reasonable, as over eóns evolutionary selection processes have forced trees to engineer themselves so that wood is a highly optimised material. A non-linear model was chosen by Barrett and Foschi to encompass failure over a broad range of stress level versus time to failure ($\sigma - T$) relationships. Their prime objection to linear $\sigma - T$ models was belief that for stress levels close to zero predicted times to failure would be unrealistically short. Thus, they like Hendrickson *et al.* (1987) adopted a threshold stress level, σ_o, below which no damage can accumulate. The Barrett and Foschi model has the form:

$$\frac{dD}{dt} = A(\sigma - \sigma_o)^B + C \times D \quad \text{if } \sigma > \sigma_o$$

$$\frac{dD}{dt} = 0 \qquad\qquad\qquad \text{if } \sigma \leq \sigma_o \qquad (8.8)$$

where A, B and C are calibration constants. In this model the rate of damage accumulation accelerates with any increase in damage. Thus, predictions from Equation (8.8) are load path dependent, i.e. recognise that wood has memory. Barrett and Foschi (1978a) also considered, but advanced argument against, a model of the form:

$$\frac{dD}{dt} = A(\sigma - \sigma_o)^B D^C \quad \text{if } \sigma > \sigma_o$$

$$\frac{dD}{dt} = 0 \qquad\qquad\qquad \text{if } \sigma \leq \sigma_o \qquad (8.9)$$

Barrett and Foschi (1978b) suggested a random variable version of their model for probabilistic treatment of static fatigue processes as might occur under the effects of sustained snow on roof or floor loads. The intention was that the random variable model be applied within Monte Carlo type structural reliability simulations. Damage threshold and current damage terms were made random variables assuming standard normal deviates, an assumption equivalent to log-normal distribution of the randomised model parameters:

$$\frac{dD}{dt} = Ae^{\varpi_1 R_N}(\sigma - \sigma_o)^B + Ce^{\varpi_2 V_N}D \quad \text{if } \sigma > \sigma_o$$

$$\frac{dD}{dt} = 0 \qquad\qquad\qquad\qquad \text{if } \sigma \leq \sigma_o \qquad (8.10)$$

where ϖ_1 and ϖ_2 are calibration constants, and R_N and V_N are random normal variates (zero median).

Foschi (1979) discusses the application of Equation (8.8) to tension tests on large capacity split-ring, nail and punched metal-plate connections in wood members loaded parallel to the grain. Focus was on methods of estimating model constants from test data, especially the 'threshold stress ratio'. For split-ring connections all failures were due to shear stress parallel to the grain. It was estimated that the threshold stress ratio was 0.55. The nailed joints were observed to fail by a combination of wood bearing and nail yielding, with a plastic hinge developed either side of the joint plane. Based on this observation, Foschi reasoned that the threshold stress ratio would be about $0.55^{1/2} = 0.742$ for nailed joints (because joint strength is proportional to root of the bearing strength of wood under nails (Johansen, 1949)). Thus, he presumed the threshold stress ratio for wood is the same for all stress states. For the toothed-plate-connectors failure was in the wood, so the threshold stress ratio was taken as 0.55. Taking the threshold stress ratio for wood governed failures as 0.55, the behaviour of nail and punched metal–plate-connection tests became verification problems. Although the data was fairly sparse, the threshold stress ratio of 0.55 fitted it quite well. Foschi's study lends credibility to the notion of a threshold stress level for static fatigue under sustained constant load. However, there is a substantial discrepancy between Foschi's estimate of the threshold stress level and the value that is implied from HCF nail plate connector tests by Hayashi *et al.* (1980). Foschi's value is reasonably close to estimates by Karacabeyli (1988) for lumber.

Foschi and Yao (1986) proposed an improved Barrett and Foschi type model that was subsequently applied within structural reliability analysis of lumber members (Foschi *et al.*, 1989). That model has the form:

$$\frac{\mathrm{d}D}{\mathrm{d}t} = A(\sigma - \sigma_o)^B + C(\sigma - \sigma_o)^N \times D \quad \text{if } \sigma > \sigma_o$$

$$\frac{\mathrm{d}D}{\mathrm{d}t} = 0 \qquad\qquad\qquad \text{if } \sigma \leq \sigma_o \qquad (8.11)$$

where N is an additional calibration constant. Karacabeyli (1988) fitted this model to extensive data for White spruce and Western hemlock lumber loaded in dry conditions under which the moisture content fluctuated between 8% and 11.5% over three years of testing. Effects of different qualities of lumber, cross-section sizes, and levels of sustained stress were investigated. Although most tests applied flexural load, there were also compression and tension parallel to grain and shear (torsion tube) tests. 'Best fit' values for the threshold stress level were quite stable, and varied between 0.42 and 0.53. As would be expected, other model parameters varied significantly between source cases.

Due to sensitivity that physical and mechanical properties of wood have to moisture, any constants fitted to empirical damage equations only apply to the 'calibration climate', which is a severe limitation on their usefulness. To overcome the limitation, Toratti (1992) further adapted the Barrett and Foschi model. Mechanosorptive damage terms were incorporated. Toratti's variant of the model takes the form:

$$\frac{\mathrm{d}D}{\mathrm{d}t} = A(\sigma - \sigma_o)^B + C(\sigma - \sigma_o)^N \times D + E(\sigma)^F \left| \frac{\mathrm{d}m}{\mathrm{d}t} \right| \qquad (8.12)$$

where E and F are calibration constant, and $\mathrm{d}m/\mathrm{d}t$ is the rate of change in moisture content. Under the model, sorption related damage accumulates only if moisture movement is coincident with the application of external load. This is a questionable presumption, as it is well known that drying alone can lead to significant damage. Damage accumulation is assumed to accumulate at the same rate under adsorption and desorption conditions, which again is questionable. Mechanosorptive behaviour is presumed to be able to activate damage accumulation below σ_o, which is a conceptual inconsistency. Aicher and Dill-Langer (1997b) argue that there should be a threshold level within the mechanosorptive term of Equation (8.12), as otherwise the mechanosorptive damage could swamp that propagated by externally applied load. Leaving aside the issue of consistent application of σ_o, it is perhaps more rational to introduce a threshold level into the dm/dt term, as very low rates of moisture movement should not cause measurable damage. Aicher and Dill-Langer combined the Foschi and Yao (1986) model with two-dimensional finite element analysis to predict static fatigue due to tension perpendicular to grain. They considered behaviour of European softwood glued-laminated-timber (glulam) specimens with a volume of 0.1 m^3. Effects of a stepped sustained loading history and cyclic variation in external relative humidity (cyclic moisture content) were accounted for within the analysis (Aicher and Dill-Langer, 1997a). Cyclic change in moisture content was taken into account using a scalar adjustment to the stress history. Agreement between theory and experiments was quite good, but practicality of such detailed yet essentially empirical analysis seems limited. It is unclear whether the approach could successfully be applied to more general situations, e.g. load histories allowing stress relaxation.

Although not the intention of their exercise, Aicher and Dill-Langer (1997b) highlighted a major weakness of all empirical damage models. Quite different combinations of model parameter led to equally good fits to the calibration data in terms of predicted times to failure, within the range of the data. This should not be surprising, given that Barrett and Foschi type models have up to five calibrated parameters (A, B, C, N and σ_o), seven if E and F are included. There must be serious concerns about application of the type of model is to load duration, load sequences or climates beyond the range of calibration data. Unfortunately, one must conclude that Barrett and Foschi type empirical models are inherently unstable and thus unreliable as extrapolation tools, which is their intended purpose!

Gerhards (1979) proposed a model for fatigue in wood and wood-based products predicated on the assumption that the residual fractional lifetime, γ, is the complement of the linear sum of fractional loading periods, t_i/T_i so that:

$$\gamma = 1.0 - \sum_{i=1}^{I} \left(\frac{t_i}{T_i} \right) = 1.0 - D \tag{8.13}$$

where t_i is loading period at stress level σ_i, T_i is lifetime at constant σ_i. D is the damage index as before ($0.0 \leq D \leq 1.0$). Replacing the summation t_i/T_i values by the integral form:

$$\gamma = 1.0 - \int_0^\tau \frac{dt}{T} = 1.0 - D \tag{8.14}$$

Equation (8.14) is analogous to the Palgrem–Miner rule (Palgrem, 1924; Miner, 1945), and when put into integral form is equivalent to Robinson's generalised model for static fatigue in metals (Robinson, 1952). If a linear relationship is presumed between σ and $log_{10}T$, the damage law becomes:

$$\gamma = 1.0 - \int_0^\tau 10^{(A-\sigma)/B} dt \tag{8.15}$$

with the rate of damage accumulation being:

$$\frac{dD}{dt} = 10^{(A-\sigma)/B} \tag{8.16}$$

where A and B are calibration constants.

The assumption of a linear σ versus $log_{10}T$ relationship is based on extensive review of test data by Gerhards (1977). Gerhards and Link (1987) proposed the following model:

$$\frac{dD}{dt} = \exp(-A + B\sigma) \tag{8.17}$$

where σ is stress level at time t normalised relative to median static strength, and A and B are calibration constants. Equations (8.16) and (8.17) are mathematically equivalent. The later was calibrated against data from tests on dry 38×89 mm Douglas fir lumber loaded in bending (49 replicates), with specimens chosen to have a dominant edge knot on the tension face. Tests were conducted with three ramp load rates until failure, and three sustained levels of stress (low, intermediate and high). Ramp loading rates produced average times to failure of 31 seconds, 127 minutes and 16.9 days. Tests with sustained stress were terminated before all specimens failed, after 4.65, 33.9 and 220 days for the low, intermediate and high stress, respectively. Surviving specimens were ramp loaded to failure after one day of recovery. When fitting the model it was assumed that model parameters are random variables, to explain variability in strength versus time to failure. Model parameters that otherwise fitted data well, did not fit the fast ramp loading data in a satisfactory manner. The 'best fit' model tended to overestimate the strength under fast ramp loading.

Figure 8.2 compares characteristic 'shapes' of above mentioned Barrett and Foschi type (Equations (8.8)–(8.11)) and Gerhard type (Equations (8.16) and (8.17)) empirical damage models, as they apply to constant sustained load. It is clear that either model can be made to fit data quite well for short to intermediate times to failure (between a few hours and several years). Available calibration data for sustained loading falls away from the right hand region of Figure 8.2, which unfortunately is where models predictions diverge significantly. To be conclusive about superiority of either type of model relative to the other, it would be necessary to collect stepped sustained load calibration data and data for quite large times to failure. A typical fit of a Barrett and Foschi type model, Equation (8.11), to data for lumber loaded in flexure is shown in Figure 8.3. The figure also shows fit of Nielsen's (1996) fracture mechanics based DVM model (see Section 8.4), and compares data and model trends to the so-called Madison curve often adopted be design codes (see Section 6.2.1).[1] It is clear that quality of a model's fit to data depends upon the stress level in question, and that models diverge significantly beyond the data range.

Cai *et al.* (2000) compared predictions from various models mentioned above to data for dry clear Southern pine subjected to flexural load sequences of up to five days duration. Models considered were Hendrickson *et al.* (1987), Equation (8.7); Barrett and Foschi (1978a), Equation (8.8); Foschi and Yao (1986), Equation (8.11); and Gerhards

---

[1] Comparison with the Madison curve is not rigorous, because that curve reflects a different interpretation of stress level when it was fit to data (not the data shown here).

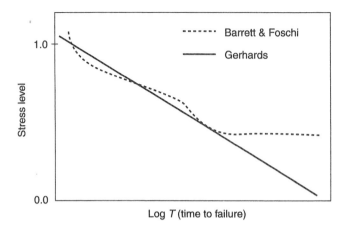

Figure 8.2   Characteristic shapes of empirical damage accumulation models of the Barrett and Foschi and Gerhard types.

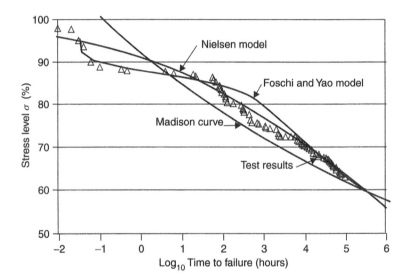

Figure 8.3   Fit of Foschi and Yao model (1986) and DVM model (Nielsen, 1996) to data for 50 × 100 mm lumber beams of Norway spruce subjected to bending at moisture content = 11%, and comparison of trends with the Madison curve (Wood, 1951). (Data from Figure 6.11; figure adapted from Hoffmeyer (2003).).

and Link (1987), Equation (8.17). Load sequences consisted of five cycles of 23 hours loading, followed by one hour of recovery (duty ratio = 0.9583, R-ratio = 0.0). In phase I of the study, the maximum stress level varied between cycles and had the sequence: 0.70, 0.83, 0.71, 0.86, 0.91, based on individually matched static control specimens (50 replicates). Phase II adopted a constant peak stress level of 0.75 or 0.80 or 0.85 or 0.90 within load cycles, or the same sequence as in phase I (8 replicates). Load sequences were designed based on average predicted bending strength of a test group, in a manner expected to maximise differences in predictions between models. Values of model constants used in predictions were those fitted by the original authors, and represented behaviour of a range of wood species, clear wood and lumber, and different loading configurations. Predictions took account of expected statistical variability in times to failure using Monte Carlo simulation. Expectations were that an 'average' specimen would fail within five load cycles. Cai *et al.* wanted to assess whether the choice of model and its associated calibration data was important with regard to predicting when their specimens would fail under five-day load sequences. For tests in phase I, all models substantially underestimated the proportion of specimens that failed during loading, and substantially overestimated the proportion of survivors after five load cycles. The proportions of failures during cycles two through five were predicted relatively well, with quite modest differences in predictions between models. In a global sense, agreement between prediction and observation was better for phase II than for phase I, but there were specific substantial discrepancies depending on the level of stress involved. None of the four models considered was judged superior to others. Cai *et al.* concluded "... that the simulated failure distributions fit the actual failure distribution reasonably well in both experimental phases". These authors differ with that judgement. Gerhards and Link (1987) found that empirical damage models overestimate strength under fast ramp loading, which fits with the finding by Cai *et al.* that the incidence of failures during loading of their specimens was underestimated.

Gong (2000) investigated compression parallel to grain Low Cycle Fatigue (LCF) of clear Norway spruce at 13% moisture content. He used a damage index of the form:

$$D = 1 - (1 - \alpha)\left(1 - \frac{(n-1)}{(N-1)}\right)^{\beta} ; \quad 1 \le n \qquad (8.18)$$

where α is damage occurring during the first load cycle, n is the number of load cycles, N is the fatigue life at the applied σ, and β is a model constant. As actual damage cannot be measured as tests progress, the relative effective modulus, E_r, was used as a surrogate for D. E_r is calculated as:

$$E_r = \frac{E_1 - E_i}{E_1} \qquad (8.19)$$

where E_1 is tangent modulus for up-loading in the first load cycle, E_i is tangent modulus for up-loading in the ith load cycle. The values of α and β appropriate to a particular σ were determined from least squares fitting data to Equation (8.18), taking $D = E_r$. The elastic modulus for the undamaged material was taken to be $E_0 = \frac{E_1}{(1-\alpha)}$. The damage index was fitted to data for square waveform loading with an R-ratio of 100 and duty ratio of 0.5 (Table 8.1 and Figure 8.4). The fatigue lives of specimens for a square waveform were compared to those for a nominally matched 'soft square' waveform (essentially a square wave with smoothed transitions between maximum and minimum stress levels (see Figure 6.4). Significantly shorter fatigue life (i.e. more rapid damage accumulation) was observed for the square waveform, even though it had a lower energy content than the soft square waveform. For compression, parallel to grain controlled failures at least, it can be concluded that both stress level and the rate of loading should appear in any generalised damage model or damage index (see Section 6.2.2).

Table 8.1   Model parameters, Equation (8.18)

Stress level, σ	α	β
0.95	0.03 615	0.03 607
0.85	0.02 967	0.02 212

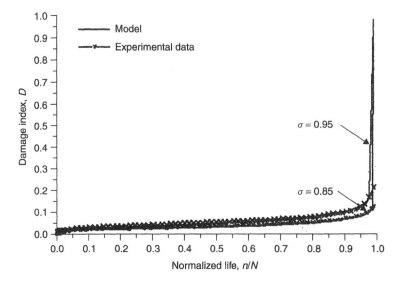

Figure 8.4   Damage models (Equation (8.18)) at two stress levels: compression parallel to grain.

Work on man-made fibre-reinforced polymer composites suggests a generalised approach to damage modelling that might be suitable for wood. Tang *et al.* (2000) proposed a cyclic fatigue model that can account for peak stress level, R-ratio, loading frequency and environmental parameters. It is assumed that:

$$D = 1 - \frac{E}{E_0}; \qquad 0 \le D \le 1 \tag{8.20}$$

$$\frac{\mathrm{d}D}{\mathrm{d}n} = \frac{C(\sigma^2(1-R))^\alpha}{(1-D)^\beta}; \quad 0 \le R; |\sigma| \le 1 \tag{8.21}$$

where E is residual modulus, E_0 is initial modulus, n is number of load cycles, σ is the peak stress level, R is the R-ratio, and C, α and β are model constants. It is suggested that C take the form:

$$C = C^*(\text{Temperature, loading frequency}, \ldots\ldots)(\sigma_{ult})^{2\alpha} \tag{8.22}$$

The model (Equations (8.20)–(8.22)) has been validated for tensile fatigue loading of pultruded vinyl ester/E-glass fibre composite, for σ from 0.35 to 0.65, R-ratio = 0.1. Loading frequencies of 2 Hz and 10 Hz, and dry air, filtered water and simulated salt-water environments at 30°C were used. The $\sigma - \log N$ curves were sensibly the same for each service environment considered (dry air, filtered water, saltwater). Possibly the same is true for wood, as is implied by Nielsen (1996). For the E-glass fibre composite tested α and β could sensibly be taken as constant, but C was a function of the loading frequency. *C* probably needs to also be a function of parameters such as peak loading rate in the case of wood.

Although individual structural wood members often exhibit essentially brittle behaviour, structural systems and connections typically exhibit ductile or pseudo-ductile behaviour, in part because they contain metal components. Thus, for systems and connections it is useful to consider how damage indices have been applied to non-wood materials (e.g. structural steel and reinforced concrete). For 'other materials' damage indices have been related to a broad range of indicating parameters (stiffness, deformation, energy dissipated), based on the relationship (Figure 8.5):

$$D = \frac{(d - d_0)^A}{(d_u - d_0)^A} \qquad (8.23)$$

where *d* is the calculated value of the damage variable (e.g. deformation), d_0 is a threshold value of *d* below which no damage accrues, d_u is the ultimate value of damage variable (e.g. deformation at failure under static load), and *A* is a calibrated exponent (taken as 1.0 in the absence of other information). An important issue is what value of the damage variable, d_r, defines the point at which damage is significant enough for repair to be necessary. Proposed damage indices involve combinations, or all of, stiffness ratio (residual stiffness divided by undamaged stiffness, dissipated energy and ductility ratio (maximum attained deformation divided by yield deformation), or related quantities. When analysing a structural system *D* can be applied at local, intermediate (sub-structure) or global levels, with values for higher levels being weighted-averages of local indices (Kappos, 1997). Kappos and Xenos (1996) used a form of the Park–Ang damage index (Park and Ang, 1985) to assess the importance of energy in combined damage indices. They considered structures of realistic size (10-storey reinforced concrete frame) with hysteresis characteristics. The form of the model employed was:

$$D = \frac{\theta_{\max}}{\theta_u} + \beta \left(\frac{\int M \, d\theta}{M_y \theta_u} \right) \qquad (8.24)$$

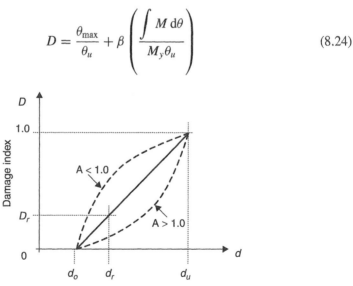

Figure 8.5   Relationship between damage variable (*d*) and damage index (*D*), Equation (8.23).

where θ is applied rotation, θ_{max} is maximum recorded rotation, θ_u is rotation capacity, M is the applied moment, M_y is yield moment, and β is a calibration constant. It was found that the contribution of the energy term to damage indices was low (up to 8.6% for beams, 5.2% for columns) when a 'typical' value for $\beta = 0.05$ was used, based on seven historical earthquake records. It is stated that for the vast majority of members the energy term accounts for 2–4% of D, and that the contribution was in some cases a maximum for those members with a relatively low D-value. If $\beta = 0.25$ is used, the maximum percentage contributions of the energy term increased by five fold. Kappos (1997) summarised results of a comprehensive comparison of indices by Fardis (1995). It is reported that most of the indices underestimate failure and the scatter associated with them is quite large. A modified version of the criteria of Park *et al.* (1987) and Fardis (1995) is reported to be the most accurate for reinforced concrete. A key difficulty is that global damage indices cannot be treated as a weighted-average of local indices. The monotonic ductility term, e.g. $\frac{\theta_{\mathrm{max}}}{\theta_u}$ in Equation (8.24), dominates damage indices for LCF of reinforced concrete structures. The same is probably true for timber structures with ductile connections. Although the above gives a sound framework for application of damage indices, it is important to be aware that design objectives and the level where indices should be applied, can differ depending upon the type of loading. Under wind load conditions, design against a local failure such as in cladding connections can be critical, because this avoids progressive and disproportionate damage in the primary structural system. Under seismic loading failure in secondary components tends to be less critical. For ductile wood systems and connections, several potentially suitable variants of Equation (8.24) (Park and Ang type models) can be envisioned.

Sletteland (1976) fitted a multiple regression equation to HCF data for punched metal-plate connector joints in lumber. For one-side of zero cyclic tensile loading, R-ratio = 0.0, the model took the form:

$$\mathrm{Log}_n N = \mathrm{Log}_n a + b\,\mathrm{Log}_n E + c\,\mathrm{Log}_n P$$

therefore $\qquad\qquad N = a E^b P^c \qquad\qquad\qquad\qquad$ (8.25)

where E is modulus of elasticity of joined members, P maximum load per tooth during load cycling, and a, b and c are calibration constants. Dependence of fatigue life N on E implies that fatigue life is a function of the density of the members, because E is proportional to density. Density of the members determines the rigidity of the 'wood foundation' on which a tooth bears. In turn, this determines the degree of curvature of the teeth, thus whether or not they will be sheared off or pull out. It is reasonable to hypothesise that a model such as Equation (8.25) will apply to nailed joints.

Overall, although quite a number of empirical damage models have been developed and applied to wood members and connections, none adequately addresses all situations and cannot reasonably be expected to have general validity. Specifically, existing models are suitable for prediction either static or cyclic fatigue, but not interaction of the two. As elucidated by Chapter 6, this is problematic because many situations exist where the interaction is important!

8.3 Phenomenological Models

Chaplain *et al.* (1996, 1998) developed a damage model for cyclic fatigue based on a Kelvin body (dashpot-and-spring) analogue. Their concept grew from observing relationships between 'flow' and relative life (n/N) for bolt embedment and joint specimens. Flow was selected as a damage index D, based on the premise that the incremental energy associated with tertiary creep corresponds to the energy consumed in micro-cracking (fatigue damage). Flow was taken equal to the deformation increment representing the difference between tertiary creep and the secondary creep trend (Figure 8.6). Incremental energy was taken to equal the sub-area under the load-deformation relationship bounded by tertiary and secondary creep trends. Normalisation of flow yielded the damage index D ($0.0 \leq D \leq 1.0$):

$$D = \frac{f_i}{f_{max}} \tag{8.26}$$

where f_i is the flow in the ith load cycle, and f_{max} is the flow at failure. Chaplain *et al.* (1998) proposed two forms of their model (Figure 8.7). These models (Equations (8.27) and (8.28)), allow cascading damage accumulation:

Chaplain et al. Model 1

$$\frac{dD}{dt} = A \left(\frac{F_D - F_o}{F_S} \right)^B + C \cdot D \quad \text{if } F_D > F_o$$

$$\frac{dD}{dt} = 0 \qquad\qquad\qquad \text{if } F_D \leq F_o \tag{8.27}$$

Chaplain et al. Model 2

$$\frac{dD}{dt} = A \left(\frac{F_D}{F_S} \right)^B + C \cdot D \tag{8.28}$$

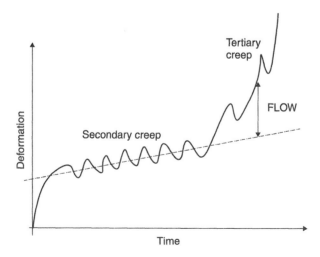

Figure 8.6 Definition of flow in models by Chaplain *et al.*

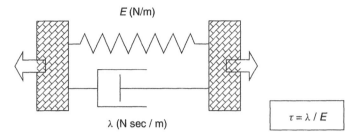

Figure 8.7   Kelvin body.

where D is the damage level at time t, F_D is the portion of the applied load in the spring of the Kelvin body (function of τ), F_o is the threshold load below which no damage accumulates, F_S is the short-term strength under monotonic loading, A, B and C are calibration constants, and τ is the ratio between traction and elastic rigidities of the Kelvin body.

Model 1 presumes there is a threshold load level below which damage does not accumulate. This model is intended to apply to either HCF or LCF analysis. Model 2 has no threshold load level and presumably is intended to only apply to LCF. Obviously, with one less parameter the latter model is simpler to calibrate. Both models were calibrated to data from bolt embedment specimens and double shear steel-wood-steel joints with one bolt. Wood members were dry spruce loaded parallel to grain. Bolts were 14 mm in diameter in a 15 mm hole. Test regimes were short-term monotonic tensile load, and sinusoidal LCF with loading frequencies of 1 Hz and 5 Hz, R-ratio = 0.1. Additional calibration tests are mentioned, but results not given in detail.

Chaplain *et al.* (1998) used their models to predict likely levels of damage in the joint of a tension bracing-diagonal in a building subjected to historical earthquake records. Comparison of predicted and observed lifetimes indicate both models (Equations (8.27) and (8.28)) are quite accurate for LCF when joints have a thin timber member, i.e. when movement of a bolt within the member approaches the behaviour of an embedment specimen. Predictions are poor for joints with slender bolts where there is significant bending deformation in the fasteners. It can be implied that the model will not perform well for joints with fasteners such as nails.

It seems feasible to confirm through physical observation whether or not incremental energy associated with tertiary creep corresponds to the energy consumed in micro-cracking, and that should be done before any widespread acceptance of the modelling concept due to Chaplain *et al.* Also, there is a need for caution in regard to use of the Kelvin body model, as it is well known that it and other simple rheological models are only capable of representing response with any accuracy over restricted time ranges and steady-state moisture conditions. It is doubtful that either model by Chaplain *et al.* can reliably predict both LCF and HCF without refinement.

8.4   Mechanics Models

Nielsen (1996) developed a Damaged Visco-elastic Material (DVM) theory for wood and other materials. DVM theory presumes damage is related to visco-elastic growth of micro cracks, and is generalised by working in terms of normalised: stress, strength, creep and time. Mode I (opening mode) behaviour of a cracked isotropic material, is the surrogate for any type of damage loci that exhibits visco-elastic growth, e.g. cracks, buckled cells. Solutions relate to a semi-infinite plate with a single crack that has a cohesive zone and plastic yielding adjacent to the tips. Normalised material compliance in the cohesive zone is described by the 'power law' creep function:

$$C(t) = 1.0 + \left(\frac{t}{\tau}\right)^b \qquad (8.29)$$

where τ is the relaxation (creep doubling) time, and b is the creep power. According to Nielsen, τ is a function of temperature and moisture content of wood, while b is independent of both temperature and moisture content. Effects of external environment on fatigue behaviour can thus be predicted by varying τ. Critical crack opening displacement, δ_c, is adopted as the failure criterion, and is related to the strength (critical stress level), σ_c, of the cracked material:

$$\sigma_c = \sqrt{\frac{E\sigma_1\delta_c}{\pi l}} = \sqrt{\frac{EG_c}{\pi l}} = \frac{K_c}{\sqrt{\pi l}} \qquad (8.30)$$

where E is modulus of elasticity, K_c is the critical stress intensity factor, l is crack length, σ_1 is the theoretical strength, and $G_c = \delta_c\, l$ is the critical energy release rate (Sections 4.3 and 4.4). The damage velocity (rate of damage accumulation) is equated to the velocity of crack growth, and for elastic cyclic fatigue the velocity of crack growth is:

$$\frac{dl}{dt} = \frac{R_{max}}{\Omega} \Rightarrow \frac{d\kappa}{dn} = \frac{\pi^2 FL^2 \kappa \sigma_{max}^2}{8\Omega/T} \qquad (8.31)$$

where FL is the strength level, n is the number of load cycles, R_{max} is the immediate crack front width, σ_{max} is the peak stress level, T is the cycle time, Ω is time at which the crack front is at position x, and $\kappa = l/l_0$ is the damage ratio. The strength level FL is the ratio of actual strength to theoretical strength of undamaged material and is related to the inherent flaw size (Nielsen, 1991):

$$FL \approx \sqrt{\frac{d}{l_0}} \qquad (8.32)$$

where d is the cell (tracheid or fibre) diameter, and l_0 is the flaw size $\approx$ longitudinal cell (tracheid or fibre) length. It can be seen from Equation (8.31) that damage accumulation is predicted to be a strong function of material quality (FL) and the stress level (σ_{max}), apart from number of load cycles and the loading frequency. Interestingly, fatigue life under a certain σ_{max} should be less for high quality material than for low quality material. This prediction contradicts the assumption that underpins the equal rank assumption (see Sections 6.2.5 and 8.5). The DVM theory distinguishes the energy required to open the crack from that needed to open and close the crack during load cycling. Lifetime can be predicted accounting for both crack closure phases of load cycles and a damage threshold. The theory has been generalised by allowing for visco-elasticity in the cohesive zone of the crack. It is predicted that visco-elastic materials always eventually fail under cyclic stress irrespective of whether or not there is a fatigue limit. Nielsen (1996) gives an iterative solution strategy, based on his theory (Appendix A of his report) for predicting damage accumulation, or lifetime, under arbitrary load histories. For damage due to pure static fatigue the time to failure or residual strength can be predicted explicitly (Equations (43) and (44) of Nielsen's report).

The DVM theory has been calibrated by Neilsen against test observations reported by other researchers. This covered LCF in compression parallel to grain, LCF in tension perpendicular to grain of specimens with artificial cracks, LCF and HCF of laminated wood, particleboards and hardboard loaded in bending, and LCF and HCF of finger jointed lumber loaded in tension parallel to grain. Figure 8.3 shows a typical fit of the DVM model to data from flexural static fatigue tests on lumber. Although the DVM model is claimed to yield good fits to data, it cannot be regarded as validated because parameters were estimated based on the test data to which the model fit was compared. Some of the parameters do not conform to expectations based on other evidence, which raises concern about physical correctness of the model.

Despite superficial complexity, Nielsen's DVM theory does not reflect the true fracture behaviour of wood (see Chapter 7). He argues, however, that this was necessary to arrive at mathematically tractable solutions. As already alluded, there is difficulty estimating material properties such as *FL*, l_0, τ and *b* that enter the model. Although available evidence indicates the theory is not yet exact or robust enough for general use, it shows significant promise and demonstrates that ultimately fracture models will be capable of predicting fatigue in wood.

8.5   Concluding Remarks and Recommendations

As will be clear from earlier discussion in this chapter, the vast majority of techniques for predicting fatigue damage in wood employ an empirical damage index. A few models have been proposed with a partially correct physical basis, but as yet these are not developed to the level where they can be calibrated and used with total confidence. For this reason, remarks below concentrate on choice of the empirical model function most likely to perform robustly in practice as an extrapolation tool.

The major objection to empirical damage models (Barrett and Foschi, 1978a, 1978b; Foschi and Yao, 1986; Gerhards, 1979; Gerhards and Link, 1987) is that most parameters have no strict physical meaning. In that case, it is very difficult to supplement models to account for factors such as moisture conditions/history or member size. Model 'constants' cannot always be calibrated in a straightforward manner and their values can be highly sensitive to presumptions that underpin the calibration exercise. For example, although not discussed in their paper, Barrett and Foschi (1978a) employ the 'equal rank' assumption while fitting their model to test data (see Section 6.2.5). As discussed in Section 8.4, the fracture mechanics based DVM theory of Nielsen (1996) predicts that damage accumulates more rapidly the higher the quality of the material. By extension that theory implies the equal rank assumption is not reliable because it is predicated on an assumption that damage tends to accumulate more rapidly the lower the quality of the material. As discussed in Section 8.2, Barrett and Foschi type models were developed in large part to mimic a typical trend in time to failure under static load as a function of stress level (Figures 8.2 and 8.3). Possibly, the trend they deduced is not real and therefore their type model inappropriate.

Accepting qualitatively at least that DVM theory is correct, any damage model must recognise that the damage accumulation rate depends upon material quality. DVM predictions regarding the influence of wood quality (*FL*) would explain reduced variability in times to failure as can be observed under static load as the σ is reduced.

Stress levels considered by various researchers have been too high, and thus times to failure too short (for strictly practical reasons), for experimentally based elucidation of which is the best form of empirical damage accumulation function. Difficulty in obtaining consistent fit of models across ramp loading rates is well documented (Gerhards and Link, 1987; Daneff, 1997; Cai *et al.*, 2000). Unpublished research at the Universite Laval, Canada under the direction of Professor Marcel Samson (deceased 1997) provides another confounding piece of evidence. Insignificant difference was observed between static and impact strengths of either solid 38 × 89 mm spruce lumber, or joints in 38 × 89 mm spruce lumber that employ punched metal plate connectors with integral teeth. Such specimens were subjected to pure moment parallel to grain, with the static load tests having a time to failure of about one minute. Failure in the joint specimens was principally due to bearing failure of wood beneath the teeth of the connectors. Although the Universite Laval studies on lumber suggest little sensitivity to the rate of loading for quite fast loading rates, this contradicts findings for clear wood loaded in compression parallel to the grain (Gong, 2000). It can be speculated that although damage at the micro to macro scales is highly sensitive to the rate of loading, massive wood members and some types of joint permit redistribution of stress as loading proceeds. It seems that inferences should not be made about rates of accumulation of measurable damage in structural components from studies on small-scale specimens subjected to nominally pure states of stress. The 'best' choice of an empirical damage model is application dependent, which explains why there is no consensus on the matter.

Currently proposed empirical damage accumulation models will overestimate residual strength of specimens surviving sustained stress (or the proportion of survivors) because it is not possible in practice to force fit of a model to be equally 'good' for all times to failure. This expectation is confirmed by the study of Cai *et al.* (2000). Specimens tested by Gerhards and Link (1987) had lower residual strength than predicted by their own damage model (Equation (8.17)). It can be deduced that the rate of damage cascades in proportion to the accumulated damage. This statement should be qualified, however, so that it is consistent with findings from other studies. The damage rate will cascade once a failure surface/path that cannot be arrested is established. It is not defensible to suppose that there exists a damage threshold stress that is independent of the load history. Such independence is contrary to fracture mechanics predictions. It has been observed for mechanical systems that variability in fatigue life depends primarily upon uncertainty about individual specimens rather than on uncertainty about details of the stochastic load history (Crandall and Mark, 1963). By analogy, uncertainty about lifetimes of wood members, or connections or systems, subjected to static or cyclic fatigue loads will depend mainly on uncertainty about the extent of the initial damage.

When selecting a damage threshold stress level σ_o, logic suggests it should apply uniformly whatever the nature of the physical process that develops stress (static loading, cyclic loading, drying, temperature and moisture gradients). The value should equal the fatigue limit as observed under repetitive cyclic loading of specimens in constant environmental conditions. In studies on metal plate-connector joints (Sletteland, 1976; Hayashi *et al.*, 1980) it has been observed that specimen failure modes can differ under static and cyclic fatigue conditions. This illustrates that reliable observation of threshold stress level can only be made using specimens subject to a pure stress state (normal or shear stress). By implication, deductions about σ_o based on behaviour of flexure specimens are not reliable. Unfortunately, the vast majority of existing data represents flexural tests, due in large part to the relative ease with which such experiments may be conducted.

Neglecting chemical and biological decomposition of the wood, it is plausible that damage will not occur provided that a certain threshold is never exceeded due to any causative agent. Any threshold stress should probably be rather lower than most estimates in the literature. As discussed in Chapter 6, wood has 'memory' and damage is load path dependent. The current threshold stress level for any piece of material is undoubtedly stress history dependent and a function of the current level of damage. This concept is supported by the report from Gerhards and Link (1987) that it is possible, although not usual, for a lumber specimen to survive a long-term sustained load but fail at a significantly lower load level when reloaded to determine its residual static strength. Leicester (1990) showed the same phenomenon based on short-term loading. Unpublished tests by Mr. Gary Daneff at the University of New Brunswick, Canada showed that axially loaded lumber members subjected to low level fully reversed cyclic load can fail after only a few cycles. The peak stress level was in the order of 0.2 of the axial strength of the members in either tension or compression. In the absence of other information, it seems reasonable to take the initial value σ_o as 0.2 (value for $D = 0$).

As shown in Chapter 6, there exists an interaction between static and cyclic fatigue effects and fatigue models need to reflect that. Also as discussed, some types of model mimic experimentally based damage functions that reflect a specific data interpretation technique (equal rank assumption). The simple Gerhards type models that employ a linear damage accumulation rate (Equation (8.16)) are preferable because expressions can be integrated explicitly.

Based on all the evidence available, the most appropriate form of empirical damage model is (Figure 8.8) ($0 \leq D \leq 1$):

$$\mathrm{d}D = \sum_{i=1}^{I} \left(\int_0^{t_i} 10^{(A-\sigma_i)/B} \mathrm{d}t_i + 1/N_i \right) \quad \text{if } \sigma_i > \sigma_{o,i} = \sigma_{o,0}(1 - D_{i-1})^C$$

$$\mathrm{d}D = 0 \qquad\qquad\qquad\qquad \text{if } \sigma_i \leq \sigma_{o,i} = \sigma_{o,0}(1 - D_{i-1})^C \quad (8.33)$$

where t_i is the period for the ith load cycle, N_i is the pure fatigue life for the peak stress level associated with the ith load cycle, σ_i is the stress ratio at time t ($0 \leq t \leq t_i$), $\sigma_{o,i}$ is the threshold stress level for the ith load cycle, $\sigma_{o,0}$ is the threshold stress level when $D = 0$, D_{i-1} is the damage level at the end of the $(i - 1)$th load cycle, and A, B and C are calibration constants. Constants A and B are calibrated based on pure static fatigue tests. In the absence on other information, it is suggested C be taken as 1.0. Determination of N_i should account for dependency on loading frequency and the waveform. When the loading frequency is low, N_i can be taken as infinity.

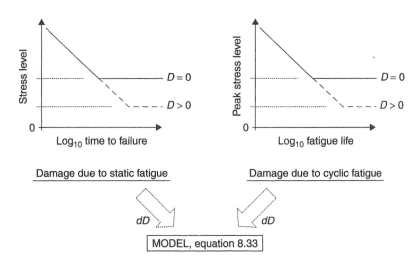

Figure 8.8 Principle of new damage accumulation model with a threshold stress level.

Equation (8.33) is based on the Palgrem–Miner rule and is an extension of Equation (8.3). For reversed load conditions the principle underpinning the model can be extended to include differences in damage accumulation rates for load applied in positive and negative senses (see Equation (8.4)). Similarly, the model can be supplemented to account for mechanosorptive effects (see Equation (8.12)). Application of the new model is demonstrated in Chapter 9.

8.6　References

Aicher, S. and Dill-Langer, G. (1997a) 'DOL effect in tension perpendicular to the grain of glulam depending on service classes and volume', Paper number 30-9-1, *Proceedings of CIB-Working Commission 18: Timber Structures, Meeting 30*, International Council for Research and Innovation in Building and Construction, Rotterdam, The Netherlands.

Aicher, S. and Dill-Langer, G. (1997b) 'Damage modelling of glulam in tension perpendicular to the grain in variable climate', Paper number 30-9-2, *Proceedings of CIB-Working Commission 18: Timber Structures, Meeting 30*, International Council for Research and Innovation in Building and Construction, Rotterdam, The Netherlands.

Barrett, J.D. and Foschi, R.O. (1978a) 'Duration of load and failure probability in wood. Part I. Modelling creep rupture', *Canadian Journal of Civil Engineering*, **5**(4): 505–514.

Barrett, J.D. and Foschi, R.O. (1978b) 'Duration of load and failure probability in wood. Part II. Constant, ramp and cyclic loading', *Canadian Journal of Civil Engineering*, **5**(4): 515–532.

Cai, Z., Rosowsky, D.V., Hunt, M.O. and Fridley, K.J. (2000) 'Comparison of actual vs. simulated failure distributions of flexural wood specimens subjected to 5-day load sequences', *Forest Products Journal*, **50**(1): 74–80.

Chaplain, M., Fournely, E. and Vergne, A. (1996) 'Wood embedment and wood joint failure prediction through a flow analysis', *Proceedings of International Wood Engineering Conference*, Louisiana State University, Baton Rouge, LA, USA: 4.54–61.

Chaplain, M., Fournely, E. and Vergne, A. (1998) 'Influence of damage models on the prediction of high stressed joint failure', *Proceedings 5th World Conference on Timber Engineering*, Presses Polytechniques et Universitaires Romandes, Lausanne, Switzerland: **1**: 313–320.

Chung, Y.S., Meyer, C. and Shinozuka, M. (1989) 'Modeling of concrete damage', *ACI Structural Journal*, **86**(3): 259–271.

Crandall, S.H. and Mark, W.D. (1963) Random Vibration in Mechanical Systems, Academic Press, New York, NY, USA.

Daneff, G. (1997) 'Response of bolted connections to pseudodynamic (cyclic) loading', MScFE thesis, University of New Brunswick, Fredericton, NB, Canada.

Fardis, M.N. (1995) 'Damage measures and failure criteria for reinforced concrete members', *Proceedings 10th European Conference on Earthquake Engineering* (1994), Balkema, Rotterdam, The Netherlands: 2.1377–1382.

Foschi, R.O. (1979) 'Load duration testing and damage accumulation modelling of timber joints', Rapport 4-79-7, Stevin-Laboratorium, Technische Hogeschool Delft, Delft, The Netherlands.

Foschi, R.O. and Yao, Z.C. (1986) 'Another look at three duration of load models', Paper 19-9-1, Volume II, *Proceedings of CIB-Working Commission 18: Timber Structures, Meeting 19*, International Council for Research and Innovation in Building and Construction, Rotterdam, The Netherlands.

Foschi, R.O., Folz, B.R. and Yao, F.Z. (1989) 'Reliability-based design of wood structures', Structural Research Series Report No. 34, University of British Columbia, Vancouver, BC, Canada.

Gerhards, C.C. (1977) 'Effect of duration and rate of loading on strength of wood and wood-based materials', US Forest Service Research Paper FPL 283, Forest Products Laboratory, Madison, WI, USA.

Gerhards, C.C. (1979) 'Time-related effects of loading on wood strength: a linear cumulative damage theory', *Wood Science*, **11**(3): 139–144.

Gerhards, C.C. and Link, C.L. (1987) 'A cumulative damage model to predict load duration characteristics of lumber', *Wood and Fiber Science*, **19**(2): 147–164.

Gong, M. (2000) 'Failure of spruce under compressive low-cycle fatigue loading parallel to grain', PhD thesis, University of New Brunswick, Fredericton, NB, Canada.

Hayashi, T., Sasaki, H. and Masuda, M. (1980) 'Fatigue properties of wood butt joints with metal plate connectors', *Forest Products Journal*, **30**(2): 49–54.

Hendrickson, E.M., Ellingwood, B. and Murphy, J. (1987) 'Limit state probabilities for wood structural members', *ASCE Journal of Structural Engineering*, **113**(1): 88–106.

Hoffmeyer, P. (2003) 'Strength under long-term loading', In: Timber Engineering, Eds. S. Thelandersson and H.J. Larsen, John Wiley & Sons, Chichester, UK.

Johansen, K.W. (1949) 'Theory of timber connectors', Publication No. 9. International Association for Bridge and Structural Engineering, Bern, Switzerland: 249–262.

Kappos, A.J. (1997) 'Seismic damage indices for reinforced concrete (RC) buildings', *Progress in Structural Engineering and Materials*, **1**(1): 1–10.

Kappos, A.J. and Xenos, A. (1996) 'A reassessment of ductility and energy-based seismic damage indices for reinforced concrete structures', *Proceedings Eurodyn'96 (3rd European Conference on Structural Dynamics)*, Balkema, Rotterdam, The Netherlands: 2.965–970.

Karacabeyli, E. (1988) 'Duration of load research for lumber in North America', *Proceedings International Conference on Timber Engineering*, Washington State University, USA: 1.380–389.

Kohara, M. and Okuyama, T. (1992) 'Mechanical responses of wood to repeated loading V: Effect of duration time and number of repetitions on the time to failure in bending', *Journal of Japanese Wood Research Society*, **38**(8): 753–758.

Miner, M.A. (1945) 'Cumulative damage in fatigue', *Journal of Applied Mechanics*, **12**(3): A159–164.

Nielsen, L.F. (1991) 'Lifetime, residual strength and quality of wood and other viscoelastic materials', *Holz als Roh-und Werkstoff*, **49**: 451–455.

Nielsen, L.F. (1996) 'Lifetime and residual strength of wood', Report: Series R, No. 6, Department of Structural Engineering and Materials, Technical University of Denmark, Lyngby, Denmark.

Palgren, A. (1924) 'Die lebensdauer von kugallagern', *Ver. Deut Ingr.*, **68**: 339–341.

Park, Y.-J. and Ang, A.H.-S. (1985) 'Mechanistic seismic damage model for reinforced concrete', *ASCE Journal of Structural Engineering*, **111**(4): 722–739.

Park, Y.-J., Ang, A.H.-S. and Wern, Y.K. (1987) 'Damage-limiting aseismic design of buildings', *Earthquake Spectra*, **3**(1): 1–25.

Robinson, E.L. (1952) 'Effect of temperature variation on the long-time rupture strength of steels', *Transactions of ASME*, **74**(5): 777–781.

Sarkani, S. (1986) 'Experimental and analytical stochastic fatigue of welded steel joints', PhD thesis, Rice University, Houston, TX, USA.

Sarkani, S. and Lutes, L.D. (1988) 'Residual stress effects in fatigue of welded joints', *ASCE Journal of Structural Engineering*, **114**(2): 462–467.

Sarkani, S., Michaelov, D.P.K and Beach, J.E. (1999) 'Stochastic fatigue damage accumulation of laminates and joints', *ASCE Journal of Structural Engineering*, **125**(12): 1423–1431.

Sletteland, N.T. (1976) 'Fatigue life of metal connector plates', MSc thesis, North Dakota State University, Fargo, ND, USA.

Tang, H.C., Nguyen, T., Chuang, T., Chin, J., Lesko, J. and Wu, H.F. (2000) 'Fatigue model for fiber-reinforced polymeric composites', *ASCE Journal of Materials in Civil Engineering*, **12**(2): 97–104.

Toratti, T. (1992) 'Creep of timber beams in variable environment', Report 31, Laboratory of Structural Engineering and Building Physics, Helsinki University of Technology, Helsinki, Finland.

Whale, L.R.J. (1988) 'Deformation characteristics of nailed or bolted joints subjected to irregular short or medium term lateral loading', PhD thesis, Polytechnic of the South Bank, London, UK.

Wood, L. (1951) 'Relation of strength of wood to duration of load', Report No. 1916, US Forest Products Laboratory, Madison, WI, USA.

Appendix: Notation

(primary or recurring items only)

$A, B \ldots$ = calibration constants
C = compliance
d = cell diameter
D = damage index
E = elastic stiffness
f = flow
FL   = strength level
HCF   = high cycle fatigue
l = crack length
LCF   = low cycle fatigue
m = moisture content
M = moment
n = number of load cycles
N = fatigue life (number of repetitive load cycles to failure)
R_N = random normal variate
R-ratio = stress ratio

t	= time/time under load
T	= lifetime (time to failure under sustained load)
V_N	= random normal variate
$\alpha, \beta \ldots$	= calibration constant
γ	= residual fractional lifetime
δ	= displacement
ω_i	= calibration constant
σ	= stress level
σ_o	= threshold stress level
τ	= creep doubling time
θ	= rotation

9

Application of Information and Concepts

9.1   Overview

There are infinite potential applications of information and concepts in earlier chapters about fracture and fatigue in wood. Use of one's knowledge can range from subjective appreciation of parameters that driver fracture and fatigue processes (leading to strategies that mitigate their influence), to application of formulae in exact engineering calculations of life expectancy or loads that will cause failure. In practice, wood is far too complex a material to allow anybody to predict its behaviour applying 'cook book' approaches, or via cold and impersonal formulae. Therefore, information and concepts should be applied as a heuristic process, with the most important ingredient for success being deep understanding of underlying mechanisms.

Analysts have often to take account of apparently contradictory and sparse data, and to be able to select an appropriate formula or model. They also have to be realistic about attainable precision. Previous chapters have approached issues such as application of models from a scientific perspective. What will be discussed now is much more practical, with emphasis on concepts rather than details of calculations because only the former need endure in the mind. No attempt has been made to be all encompassing, because that is an impossible task. Problems presented below illustrate some situations in which current knowledge can be applied with confidence. They are discursive in the nature and it is hoped they demonstrate the need for a marriage of knowledge of physical processes with modelling capabilities.

9.2   Problem 1: Brittle Failure in Multiple Bolt Joints

This problem illustrates prediction of brittle failure of a component that incorporates wood that is initially free of cracks. Consideration is given to an axially loaded 4-bolt double-shear steel-glulam-steel splice joint (Figure 9.1). Steel bolts are located within

Fracture and Fatigue in Wood   I. Smith, E. Landis and M. Gong
© 2003 John Wiley & Sons, Ltd   ISBN: 0-471-48708-2 (HB)

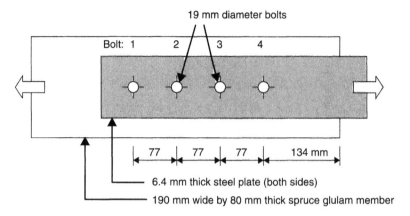

Figure 9.1 Problem 1: Arrangement of four bolt double-shear steel-glulam-steel joint.

holes with a 1.6 mm tolerance, and have low slenderness within the thickness of the central member so exhibit negligible bending deformation. Load-slip plots for such joints imply almost perfect brittle-elastic response under monotonically increasing load that causes failure in about 0.1 hours. However, in the purely elastic regime, there is distinctly unequal sharing of load between bolts because of deformation in the wood member (steel side plates are rigid enough that their extension can be ignored). For the particular arrangement, the bolt nearest the end of the wood member (bolt 4) transfers the least load between members, and the bolt furthest from the end (bolt 1) transfers the most load. Failure of the joint occurs in the central wood (glulam) member and results in 'shearing-out' of blocks of material from between adjacent bolts and from between the last bolt and end of the member (Figure 9.2). Even superficially, it is clear that failure is ultimately due to excessive shear stress parallel to grain.

If isolated bolts (single bolt joints) are tested the load-slip response is almost perfectly elasto-plastic, and as shown in Figure 9.3. After a certain amount of plastic slip, termed the Plastic Slip Limit (PSL) in this discussion, there is sudden loss of strength due to unstable crack growth. Within the plastic regime the mechanism by which load is transferred from wood to a bolt is quite complex, as illustrated in Figure 9.4. Immediately under the bolt there is a 'stick zone', where stress normal to the bolt surface is

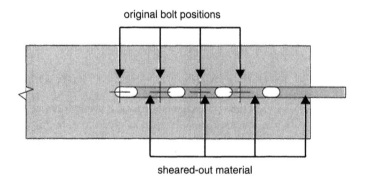

Figure 9.2 Problem 1: Row shear-out failure in glulam member.

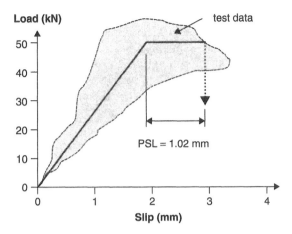

Figure 9.3 Problem 1: Load-slip response of an isolated bolt, and definition of Plastic Slip Limit (PSL).

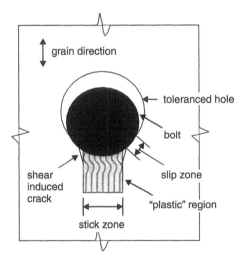

Figure 9.4 Problem 1: Idealisation of load carrying mechanism for an isolated bolt (in glulam member) prior to attainment of PSL.

large enough to frictionally restrain the wood from sliding. Within this zone the wood fails in compression parallel to grain, and because of its confinement the response mimics plastic behaviour (as reflected in the bolt load-slip response in Figure 9.3). Either side of the stick zone, there is a 'slip zone' within which there is wood to bolt sliding-contact. The wood to bolt contact area is a nonlinear function of load level, with sizes of stick and slip zones depending on the ratio of bolt diameter to hole diameter and roughness of the bolt surface. A short crack develops parallel to grain at each boundary between stick and slip zones. This is mainly due to excessive shear parallel to grain stresses. Initially, these cracks are stable, but they are loci for unstable failure once the PSL is reached.

The above behaviour has been analysed by Finite Element (FE) analysis. Several FE approaches are possible, with the most general being a continuum mechanics representation that account for geometric nonlinearity at wood to bolt contact surfaces and elasto-plastic compressive behaviour in the stick zone (Kharouf, 2001, Section 3.4.3). Because bolts do not bend, it is possible to use two dimensional plane stress elastic analysis of the wood member as a whole, with account of geometric nonlinearity at wood to bolt contact surfaces. As a simple alternative to generalised elasto-plastic analysis, a special rigid-plastic element located immediately below the bolt can mimic the plastic deformation (Tan and Smith, 1999), which is the approach underpinning results discussed below. The analysis ignored deformation in steel side members, enabling zero movement of bolt surfaces to be applied boundary constraints. Load was applied to the wood member remote from the joint region. Output was load-slip responses for the individual bolts and the joint system (Figure 9.5). As shown in the figure, analysis predicts that at very large joint slip all bolts transfer the same load (50 kN per bolt). However, it is known that the plastic slip is limited by the PSL.[1] Thus, the PSL was applied as a limit on the plastic slip of each bolt. Once any bolt achieved the PSL its load was reduced to zero (total force redistribution to other bolts). In the case of the 4-bolt system considered here, load redistribution could not be sustained once slip of bolt 1 reached the PSL. This led to unzipping of the system. At that stage bolt 2 had almost reached its potential capacity, but bolts 3 and 4 only carried about half their potential capacities. The predicted joint capacity was 144 kN, which compared with an experimentally derived capacity of 138 kN (average for 10 replicates). Similar levels of agreement have been found for other multiple bolted joint arrangements (Tan and Smith, 1999).

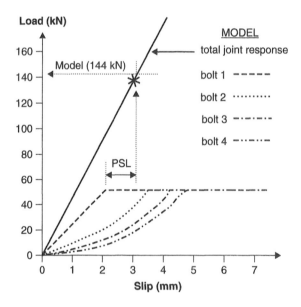

Figure 9.5   Problem 1: Predicted load-slip response for each bolt and the joint (✳ indicates the average failure load in tests = 138 kN).

---

[1] This approach negates the need to model initiation and propagation of cracks, and is therefore very computationally efficient.

This problem illustrates that there is a direct relationship between failure processes in single and multiple bolt connections. More generally, it shows that it is possible to predict ultimate load capacities of quite complex systems that fail in a brittle manner, if appropriate analytical techniques are employed.

9.3   Problem 2: Continuum Failure Models Applied to Slotted Plates

Various models that presume an uncracked body have been advanced as means of predicting failure loads for components made of brittle material. This problem examines how well four such models perform in the context of a rectangular wood plate loaded in tension and having a slot with a rounded end cut normal to one edge (Figure 9.6). The grain of the wood runs parallel to the axis of the slot resulting in a failure plane that lies directly ahead of the slot. This problem is chosen because tension perpendicular to grain is the condition that renders wood's behaviour closest to perfectly brittle-elastic. The configuration is geometrically unstable, and it is assumed in analysis that all components of stress other than tension perpendicular to grain have negligible influence on the failure mode.

Tests were carried out on 8 mm thick spruce plates with a 'slot end radius', r, ranging from 0.67–53.7 mm.[2] Specimens with $r = \infty$ were also tested (i.e. tension coupons 35 mm wide). Oven dry density of plate material was 390–480 kg/m^3, and the moisture content 15%. Average mechanical properties were: modulus of elasticity parallel to grain $E_x = 11.6$ GPa, modulus of elasticity perpendicular to grain $E_y = 0.6$ GPa, in-plane shear modulus $G_{xy} = 0.8$ GPa, Poisson's ratio $v_{xy} = 0.37$, and tensile strength

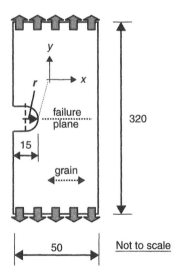

Figure 9.6   Problem 2: Geometry of slotted plate ($r = 0.67, 2.17, 3.39, 8.2, 53.7$ mm, ∞).

---

[2] Test data, and predicted elastic stress distributions used as the basis of predictions in this problem, were originally reported in the paper by Tan and Smith (1998).

$f_t = 5.47$ MPa (corresponds to plate strength when $r = \infty$). There were between 6 and 10 replicates per slot end radius. Ends of plates were held by specially designed grips that prevented in-plane rotation along those boundaries. Loading rates were such that specimens failed about two minutes after the start of a test. Failures were macroscopically brittle and catastrophic. Elastic stress distributions were predicted by finite element approximation for each slot end radius using average elastic properties and assuming a plane stress condition. Four node isoparametric elements were employed with a very fine mesh adjacent to the slot end and along the ligament ahead of it. The ligament corresponds to the failure plane. Figure 9.7 illustrates distributions of tension stress perpendicular to grain, σ_y, along the ligament for various values of r.

Considering only σ_y, failure models examined are:

• Maximum stress criterion: this is the simplest possible model and takes the form:

$$\sigma_y = f_t \tag{9.1}$$

The criterion presumes the structure of the material allows no localised toughening, i.e. it is perfectly brittle-elastic. The maximum tensile stress always occurs at the tip of the slot, which is the location presumed to initiate a cascading failure.

• Point stress criterion: this criterion assumes that failure occurs when the stress at a characteristic distance, d_o, from a stress discontinuity equals the uniaxial strength of the material. The dimension d_o is assumed to be a material property and can be thought of as the distance over which the material must be critically stressed to activate a flaw. The model takes the form:

$$\sigma_y(x = d_o) = f_t \tag{9.2}$$

Dimension x is measure from the slot tip along the ligament. Based on results for $0.67 \leq r \leq 53.7$ mm it was found that $0.4 \leq d_o \leq 0.8$ mm. The median value of $d_o = 0.6$ mm is presumed to be the best estimate for the purposes of subsequent analysis and discussion.

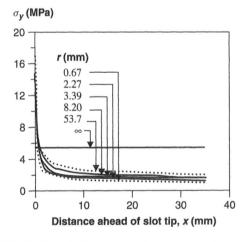

Figure 9.7 Problem 2: Stress normal to fracture plane, σ_y (MPa).

- Average stress criterion: this criterion assumes that failure occurs when the average stress over a characteristic distance, a_o, equals the uniaxial strength of the material. The dimension a_o is assumed to be a material property, and represents the distance over which the material can redistribute stress due to localised pliability. The form of the model is:

$$\frac{1}{a_o} \int_0^{a_o} \sigma_y \, \mathrm{d}x = f_t \qquad (9.3)$$

This form of the model applies under plane stress or plane strain conditions only, and assuming that the location of the failure surface is known *a priori*, as in the current problem. Based on results for $0.67 \leq r \leq 53.7$ mm, it was found that $1.5 \leq a_o \leq 2.8$ mm. The value of $a_o = 2.0$ mm is presumed the best estimate for the purposes of subsequent analysis and discussion.

- Weibull theory: like the maximum stress criterion, the theory presumes the most critically stressed 'unit(s)' of material within a continuum trigger failure. However, Weibull theory places prediction of which unit(s) triggers failure within a statistical framework (see Section 7.2). Based on Equations (7.6) and (7.8), and presuming a two-parameter Weibull model, the probability of failure of a plate is:

$$p_f = 1 - e^{-\frac{1}{A_o} \int_A p_o(x) \, \mathrm{d}A} \qquad (9.4)$$

where $p_o(x) = \left(\frac{\sigma_y}{\omega}\right)^m$, A is the area of the ligament (failure plane), A_o is the representative area element ($=280$ mm^2), ω is the scale parameter ($=5.93$ MPa), and m is the shape parameter ($=5.0$). Quoted values of A_o, ω and m were calibrated based on the results from tension coupons ($r = \infty$). Equation (9.4) is put in terms of an area (rather than volume) integration for this problem because the location of the failure plane is known.

Table 9.1 summarises average test capacities of the slotted plates, and ratios of predicted to observed capacities for each of the above mentioned models. The maximum stress criterion is the most sensitive to the stress concentration adjacent to the slot tip, and always yields the lower bound estimate of failure load for a plate. As can be seen from the table, the degree of conservatism in predictions increases if r is decreased. In the extreme the plate capacity would be predicted to be zero at a slot tip radius of zero, i.e. $\sigma_y(x = 0) = \infty$ for an infinitely sharp crack. Point stress and average stress criteria both yield conservative and non-conservative predictions of plate capacities, with the average stress criterion being the more consistently accurate. It should be noted however that tabulated ratios 'model prediction/failure load in test' reflect the specific choices of characteristic dimensions d_o and a_o. In turn, d_o and a_o reflect the database on which they were calibrated. It can be claimed justifiably that, compared with other criteria, the average stress criterion is consistently reliable as a basis for predicting capacities of slotted plates. The same is true for other components where tension perpendicular to grain is the dominant concern.

Table 9.1 Problem 2: Summary of test and model predictions

Slot tip radius r (mm)	Av. failure load in tests (N)	Model prediction/Failure load in test			
		Maximum stress	Point Stress	Average stress	Weibull theory
0.0	not tested	0.00	→0	→0	→0
0.67	463	0.32	1.28	1.14	1.11
2.17	498	0.37	1.12	1.09	1.18
3.39	526	0.39	1.02	1.04	1.25
8.2	658	0.43	0.83	0.88	1.25
53.7	758	0.83	0.95	1.03	1.63
∞	1,532	1.0	1.0	1.0	1.0

A practical problem applying the average stress criterion is that there is no direct means by which to measure a_o in a test. However, Landelius (1989) suggested estimation of the dimension from the relationship:

$$a_o = \frac{2E_y G_f}{\pi f_t^2} \tag{9.5}$$

where G_f is the specific fracture energy in mode I. Taking $G_f = 260$ J/m^2 (Table 5.3), and other properties as above, yields an estimated $a_o = 3.32$ mm. This compares reasonably well with the range of values ($1.5 \leq a_o \leq 2.8$ mm) that would align the model predictions exactly with individual results for plates with various r. Had it been assumed in the current problem that $a_o = 3.32$ mm instead of 2.0 mm, predicted capacities would have been non-conservative by between about 10 and 40%. It should be borne in mind when evaluating Equation (9.5) that G_f used here is an estimated value, and error in that could easily account for discrepancy in estimates of a_o. As is self-evident, error in a_o is most critical when stress gradients are severe.

Weibull theory gives markedly non-conservative predictions of plate capacities, except when r is quite small. It should not be presumed, however, that predictions will be meaningful when r is very small. As illustrated in Figure 9.8, when the slot tip is infinitely sharp ($r = 0$) the Weibull capacity goes to zero because, owing to infinite stress, the probability of failure approaches infinity for the volume increment located at $x = 0$. The reason why the theory is always non-conservative in the current problem for $0.67 \leq r \leq 53.7$ mm is that the stress analysis underestimates the stresses near the tip of the slot, and thus underestimates the probability of failure according to Equation (9.4). This reflects an error in the assumption that the plate is a continuum with smooth boundaries and not the analysis itself. As discussed in Section 2.2, even when there are no visible imperfections in wood it contains inherent damage in the form of microscopic cracks. Also, it is impossible to machine a slot in wood without causing damage to the machined surfaces (HMSO, 1955). Even when the greatest care is taken during manufacture of plates, there are short cracks (possibly ≪ 1 mm long) that significantly affect the high stress region around the tips of slots. When the presence of such cracks is accounted for in an analysis, stress levels (σ_y) are raised locally. Unfortunately, doing this does not solve the problem with regard to application

Weibull / Test capacity

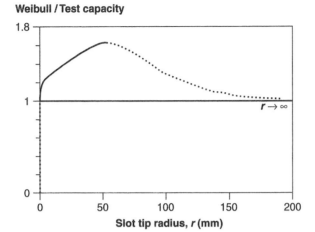

Figure 9.8   Problem 2: Dependence of the ratio of predicted Weibull capacity to observed capacity on slot tip ratio.

of Weibull theory because the crack(s) causes a stress singularity.[3,4] Although perhaps counter intuitive, the error from neglect of inherent or machining damage on stresses effects the outcome of Weibull predictions most for the larger slot end radii (within a finite range of r). In the context of slotted plate problems, the theory is only strictly reliable when r approaches infinity and stress gradients are relatively mild.

In summary: for plates with slots having rounded ends, and when tension perpendicular to grain dominates the failure process, the average stress criterion is the most reliable basis for prediction of plate capacity when assuming that the body is an elastic continuum. This approach is reliable as long as the slot tip radius does not approach zero. For very sharp tipped slots (natural cracks, artificial sharp cracks, square tipped notches), it is necessary to adopt fracture mechanics concepts for reliable prediction of plate capacities.

9.4   Problem 3: Critical Load Levels for End-Notched Members

Design of bending members notched on the tension face is one of few widely accepted applications of Linear Elastic Fracture Mechanics (LEFM) to wood. Re-entrant corners of notches, such as those shown in Figure 9.9, are loci for stress concentrations, and hence places from which cracks emanate once applied load reaches a critical level. As wood members are inherently strongly anisotropic, cracks develop parallel to the grain

---

[3] Neglect of inherent or machining damage on stresses is less important when applying either the point stress or average stress criterion, because the incremental stresses that are generated ahead of the slot are of the $x^{-1/2}$ type and therefore quickly damp out.

[4] A stress singularity is always produced under an assumption of linear-elastic material response and LEFM principles apply.

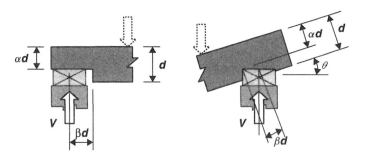

Figure 9.9   Problem 3: Typical end-notch details for wood members.

direction whatever the notch geometry. Tests have been performed at the University of New Brunswick on end-notched lumber and glued-laminated-timber (glulam) members made of spruce. Lumber specimens were 35 or 38 mm thick by between 60 and 235 mm deep, while glulam specimens were 80 mm thick by 228 mm deep. Members had simple end supports and spans about 10 times member depth. Both flat and inclined arrangements were tested to create notches of the types illustrated in Figure 9.9. Loading was displacement controlled to initiate cracking in about two minutes, and cause complete failure in about five minutes. Material properties including fracture toughness K_{Ic} were measured. Discussion here is focused on LEFM predictions of critical load levels for dry material.[5]

Stress intensity factors K_I (opening mode) and K_{II} (forward shearing mode) were calculated by finite element analysis, assuming linear elastic transversely orthotropic material response. The longitudinal direction of material symmetry was assumed to lie parallel to a member's axis. Analysis presumed an initial crack 0.1 mm long and parallel to the member axis at the re-entrant notch corner. This produced an $r^{-1/2}$ stress singularity. Based on various proposed mixed mode failure criteria (Valentin *et al.*, 1991) the mode II component contributed negligibly to the estimation of critical reaction force V_c for each type of member analysed. Therefore, the assumed failure criterion was $K_I = K_{Ic}$.

In an overall sense, LEFM accurately predicts reaction forces V_c for crack initiation in notched members, (Figure 9.10). The theory predicts that end notched members are inherently unstable with respect to crack propagation, i.e. $V_c = V_u$ (ultimate reaction force). However, when LEFM solutions for V_c are compared to experimental observations of V_u, predictions are found to be quite conservative, (Figure 9.11). This reflects that knots, and finger joints in the case of glulam, can be very effective toughening and crack arrest features. The discrepancy illustrated by Figure 9.11 is mainly due to inability of the theory to account for the anisotropic and inhomogeneous nature of lumber and glulam or toughening. It can be deduced that LEFM predictions will tend to be conservative for massive wood members and systems.

Closed form approximate LEFM equations for notched wood members have been developed and are widely used in design. For example, provisions in the Canada design code for wood are based on the following simple expression (Smith and Springer, 1993;

---

[5] Findings are similar for green material.

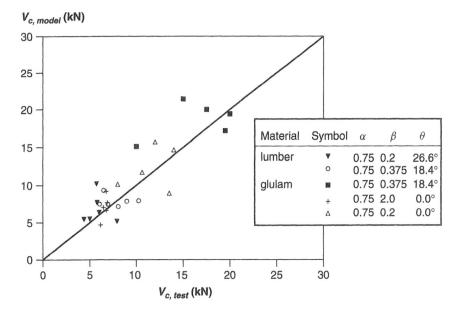

Figure 9.10   Problem 3: Predicted crack initiation reaction forces $V_{c,\text{model}}$ versus observed crack initiation reaction forces $V_{c,\text{test}}$.

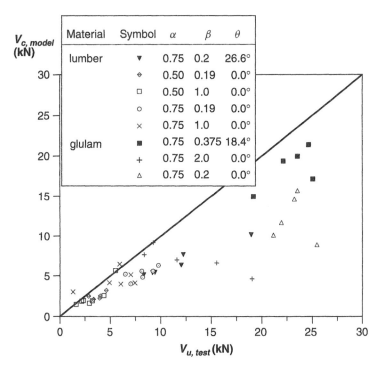

Figure 9.11   Problem 3: Predicted crack initiation reaction force $V_{c,\text{model}}$ versus observed ultimate reaction force $V_{u,\text{test}}$.

Smith *et al.*, 1996a, 1996b):

$$V_c = \frac{b\alpha\sqrt{G_c\,\mathrm{d}}}{\sqrt{\dfrac{0.6}{G_{xy}}(\alpha - \alpha^2) + \dfrac{6\beta^2}{E_x}\left(\dfrac{1}{\alpha} - \alpha^2\right)}} \tag{9.6}$$

where d, α and β are as defined in Figure 9.9, b is thickness of the member, G_c is mode I critical elastic energy release rate, G_{xy} is shear modulus in the plane of the member, and E_x is modulus of elasticity parallel to grain. When Equation (9.6) is applied to an inclined member, V_c is the component of the reaction force normal to the member axis. The equation gives accurate solutions when notches are long and/or deep, but accuracy can decrease sharply when notches are short and shallow.

In general, LEFM analysis of wood works well in situations where the stress intensity region adjacent to a notch, or any other type of stress raising feature, is small compared to general dimensions of a member. Structural arrangements must produce crack tip strain fields that are sensibly the same as those in test specimens from which fracture toughness is characterised. Provided these conditions are met, it does not actually matter whether or not fracture behaviour is truly linear or elastic. What does matters is that material behaviour is treated consistently when analysing material property test data, and when analysing structural members.

Apart from members notched on the tension face, requirements for successful application of LEFM are met by bending members with large holes (Aicher *et al.*, 1995), but in that case mode II rather than mode I fracture is the dominant influence in the failure criterion.

9.5   Problem 4: Critical Crack Length in a Reinforced Glulam Girder

'Damage tolerant design' is a problem widely studied in the aerospace industry, but to-date it has had limited application in civil engineering. This concept marries the fields of fracture mechanics and non-destructive evaluation to examine the relationships between crack sizes and structural safety. This relationship is coupled with a prescribed periodic inspection schedule designed to detect cracks that might compromise the safety of a structure before they have a chance to cause failure (Bray and Stanley, 1989). Typically, critical crack size is established for a particular component under a given loading using the fracture mechanics concepts of Chapter 4. As long as any cracks remain smaller than the critical size, the component is safe to remain in service. However, fatigue or other environmental conditions may cause cracks to gradually increase in size, potentially approaching the critical size.[6] At that time the component must be removed from service. The regular inspection interval is dictated by how fast cracks grow, and is generally set so that there will be multiple chances to identify a crack before it attains critical size.

An example of a critical crack analysis was conducted for a glulam bridge girder reinforced with a Fibre Reinforced Polymer (FRP) composite bonded to the tension

---

[6] See the discussion of sub-critical crack growth in Section 4.3.5.

side of the girder. The reinforcing technology allows low-grade wood to be used in structural applications. However, for the technology to work, the bond between the wood and the FRP must have sufficient toughness to resist the shear stresses that arise at the bond line between the two materials.

An example of a reinforced glulam section is shown in Figure 9.12. In this study, a two dimensional finite element model was created for the beam, with the wood, FRP, and bond line each represented by an appropriate orthotropic constitutive relationship. The mesh for a beam with a delamination starting at the end of the bondline is shown in Figure 9.13. In this particular example, the wood is Douglas fir, and the reinforcement is unidirectional pultruded glass fibre in a phenolic matrix.

There are several ways to determine a critical crack length for a given load. In this case, the load is due to a large heavily loaded logging truck. The most common approach is to determine the stress intensity factor, K, or the strain energy release rate, G, for the particular geometry. However, this particular problem was complicated by the fact that it is a mixed mode stress state at the crack tip, with two different materials contributing to the energy release rate. Thus, a J-integral approach was used.

In Section 4.4.4, the J-integral is introduced as a path-independent line integral of the form:

$$J = \int_{\Gamma} \left(w\, \mathrm{d}y - \mathbf{T}\frac{\partial \mathbf{u}}{\partial x}\, \mathrm{d}s \right) \tag{9.7}$$

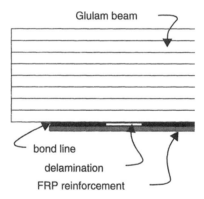

Figure 9.12   Problem 4: End portion of a reinforced glulam beam.

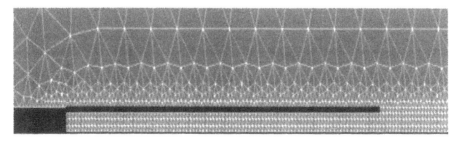

Figure 9.13   Problem 4: Finite element mesh showing debonded wood-FRP joint.

where Γ is the contour of integration, $\mathbf{T}$ is the stress vector acting perpendicular to the contour, $\mathbf{u}$ is the displacement vector, and w is the strain energy density, given by:

$$w = \int_0^\varepsilon \sigma(\varepsilon)\,d\varepsilon \qquad (9.8)$$

The quantity J is analogous to the energy release rate, G, in that it represents the energy available to drive a crack. However, J can be applied to nonlinear elastic materials, which both wood and FRP can be approximated to be so long as there is not significant unloading.

Evaluation of J can be done numerically within the finite element framework. A contour is established around the crack tip, as shown in Figure 9.14. Strain energy, displacement and traction can be determined at a number of points on the contour, and J can be integrated numerically. It should be noted that J is a path-independent quantity, so the choice of integration contour should be arbitrary. However, it was found that contours that are very close to the crack tip suffer from numerical artifacts likely resulting from insufficient integration points (Sanchez, 2002).

By varying the length of the debonded section, the influence of crack length on J can be evaluated. An example of this influence is illustrated in Figure 9.15. As

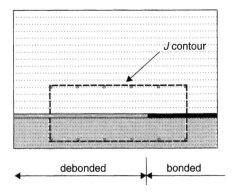

Figure 9.14   Problem 4: Contour around crack tip used for numerical evaluation of J.

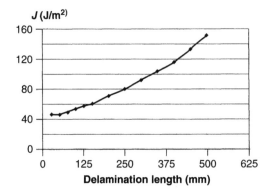

Figure 9.15   Problem 4: Calculated variation of J with delamination length.

one would expect, the energy available to grow the crack increases with crack length. To determine the critical crack length, these values must be compared with known values of fracture toughness or fracture energy. In this particular example, results from laboratory fracture energy tests of different FRP formulations bonded to Douglas fir ranged from 200–400 J/m^2, indicating that the crack must be considerably longer than the 508 mm (20 inches) examined.

Clearly, such information is useful to a bridge inspector evaluating the load capacity of a particular bridge. Periodic inspection data would likely produce a range of crack growth rates that could, in turn, be used to establish a regular inspection and maintenance program.

9.6   Problem 5: Static Fatigue in Roof Joists

This problem illustrates the application of damage models to predict static fatigue caused by variable load. Consideration is given to the behaviour of repetitive parallel joist arrangements in flat roofs that experience effects of heavy snow load. It is assumed roofs have 38 × 235 mm joists of Spruce-Pine-Fir, No. 2 grade, spaced at 400 mm on centre, and that these are sheathed over with plywood attached by nailing.[7] For that type and size of lumber, the 5-percentile static moment capacity is 5.4 kNm. The 'maximum span' for such joists that building regulations permit is about 4.1 m for domestic construction. For the purpose of the analysis, joists are assumed simply supported at each end and subject to only dead and snow load (Figure 9.16).[8] Snow loading is simulated based on conditions for Ottawa, Canada and assuming an Extreme Type I distribution. The simulated dead load is 0.5 kPa, and snow load pulses have a mean value of 1.09 kPa (CoV = 0.935) with five random length pulses per year over a November–March winter period. Simulated stress levels σ range from just above 0.0 to over 0.8, depending on joist span ($\sigma \propto$ span^2). Details of snow load simulation techniques can be found elsewhere (Foschi *et al.*, 1989).

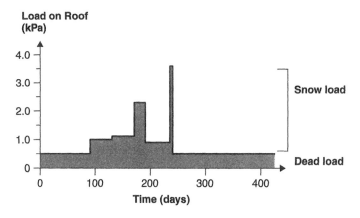

Figure 9.16   Problem 5: Segment of simulated load history for flat roofs in Ottawa.

---

[7] Nailing insufficient to enforce composite action between joists and sheathing.
[8] Typically, suction wind load on roofs is of short duration, and is inadequate to cause damage inducing stress reversals.

As discussed extensively in Chapter 8, there are various empirical models for predicting accumulation of damage (damage proportional to index D, ($0 \le D \le 1.0 =$ failure)). Calculations summarised below are based on Equation (8.33):

$$dD = \sum_{i=1}^{I} \left(\int_{0}^{t_i} 10^{(A-\sigma_i)/B} \, dt_i + \frac{1}{N_i} \right) \quad \text{if } \sigma_i > \sigma_{o,i} = \sigma_{o,0}(1 - D_{i-1})^C$$

$$dD = 0 \qquad\qquad\qquad \text{if } \sigma_i \le \sigma_{o,i} = \sigma_{o,0}(1 - D_{i-1})^C \quad (9.9)$$

where t_i is the period for the ith load pulse, N_i is the pure fatigue life for the peak stress level associated with the ith load pulse, σ_i is the stress ratio at time t ($0 \le t \le t_i$), $\sigma_{o,i}$ is the threshold stress level for the ith load pulse, $\sigma_{o,0}$ is the threshold stress level when $D = 0$, D_{i-1} is the damage level at the end of the $(i-1)$th load pulse, and A, B and C are calibration constants. For this problem, the cyclic fatigue effect does not need to be considered as snow accumulates slowly, i.e. any rate of loading effects on damage are negligible. Thus, the model reduces to:

$$dD = \sum_{i=1}^{I} \int_{0}^{t_i} 10^{(A-\sigma_i)/B} \, dt_i \quad \text{if } \sigma_i > \sigma_{o,i} = \sigma_{o,0}(1 - D_{i-1})^C$$

$$dD = 0 \qquad\qquad\qquad \text{if } \sigma_i \le \sigma_{o,i} = \sigma_{o,0}(1 - D_{i-1})^C \quad (9.10)$$

Here constants A and B are taken to be 0.9 and -0.05, respectively, based on the data shown in Figure 6.12. Parameters $\sigma_{o,0}$ and C that control evolution of the threshold stress level ($\sigma_{o,i}$) are taken to be 0.2 and 1.0, respectively. However, as such negligible damage is predicted for periods when σ_i is less than about 0.4, results would be unaffected if it were assumed $\sigma_{o,0} = 0.0$. In practice there would be random variation between joists in parameters controlling damage accumulation rates (A, B, C and $\sigma_{o,0}$), but that is ignored here.

Because joist strengths vary both within and between roofs, stress levels, and therefore damage accumulation and/or likelihood of failure, is a stochastic process. Accumulation of D is illustrated in Figure 9.17 for a joist that corresponded to the lower 5-percentile strength level (when new), as a function of span and time under load. As can be seen, such a joist should experience no appreciable damage if used within its accepted span range (≤ 4.1 m), but could acquire significant damage if the span were longer. It follows that some very weak joists can be expected to fail even if used at the permitted span. Most joists are substantially under-utilised, from a strength point of view. Table 9.2 indicates how time to failure varies with joist span and gives estimates of annual probability of failure of roof joists (per joist per year basis). As will be deduced from this problem, a primary application of static fatigue models is as tools within structural reliability analyses (see Problem 6, Section 9.7).

Quite often, damage accumulation models, such as Equation (9.10), are used to predict residual static strength following a period of in-service loading. In such instances, models are used in conjunction with a monotonic loading function that causes failure in about 0.1 hour. Such an application of models is not attempted here, because predictions are unrealistic and imply only slight reduction in strength, even when about half

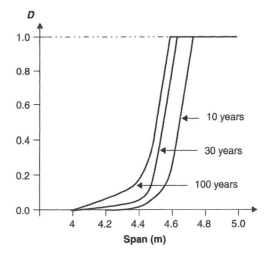

Figure 9.17   Problem 5: Effects of joist span and time in service on damage index D.

Table 9.2   Problem 5: Effect of span on time to failure and probability of failure

Span (m)	Approx. time to failure of 5% of joists (years)	Approx. annual probability of failure (per joist)
4.0	∞	10^{-6}
4.2	2×10^{3}	10^{-5}
4.4	100	10^{-4}
4.6	<10	10^{-2}
4.8	<5	10^{-1}
5.0	<1	1.0

the fractional lifetime γ has been consumed ($\gamma = 1.0 - D$). The inability of index-based damage accumulation models to perform this latter task reflects that they cannot be made to fit all times to failure (T = pure static fatigue lifetimes) equally well (see Section 8.5).[9]

It needs to be emphasised that above-mentioned results, and results of similar studies, are highly dependent on model constants, especially A and B. Often there is a high degree of uncertainty about suitable values of those parameters. Therefore, any static fatigue predictions for wood yield only 'hand waving' estimates of damage or failure rates. Results always need to be used in conjunction with engineering judgement.

9.7   Problem 6: Static Fatigue in Bolted Truss Joints

This problem considers application of damage models within Monte Carlo type structural reliability simulations. The study presumes the capacity of trusses is controlled by

---

[9] One solution to the problem is piecewise calibration of model constants, but to be reliable that requires sufficient data to determine their dependence on loading rate.

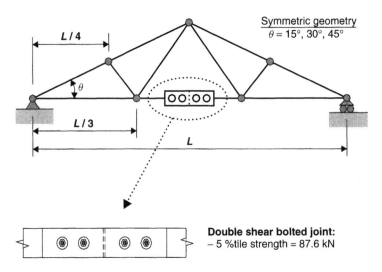

Double shear bolted joint:
− 5 %tile strength = 87.6 kN

Figure 9.18   Problem 6: W-truss with bolted splice joint in the tie member.

the bolted splice-joint located within the bottom (tension) chord of each of the trusses (Figure 9.18). Short-term (static) joint strength is taken to have a three-parameter Weibull distribution (Hanh and Shapiro, 1967) with shape parameter = 3.03, scale parameter = 25.4 kN and location parameter = 78.1 kN. Combined influences of dead and snow load are considered. The dead load on the rafter members is taken to be 0.5 or 0.75 kPa (CoV = 0.10) to represents roofs with either light or heavy roof tiles, while dead load on tie members is taken to be 0.4 kPa (CoV = 0.2). Snow loading is for light and heavy snowfall locations in the United Kingdom. Because long-term records exist for depth of snow on the ground, reliability simulations use actual snow histories, rather than simulated ones as in Problem 5.[10] For each combination of variables, at least 10 000 conceptual trusses are generated with randomised combinations of initial strength, load history and parameters controlling the rate at which damage accumulates under the effects of applied loads.

For each truss, damage in its tension member joint is presumed to accumulate according to Gerhards' model (Gerhards, 1979):

$$\frac{dD}{dt} = 10^{(A-\sigma)/B}$$

where D is a damage index ($0 \leq D \leq 1$ = failure), and A and B are calibration constants. Figure 9.19 shows two Gerhards curves of time to failure under sustained load of a certain stress level. The curve representing $A = 0.93333$ and $B = -0.06667$ is a relatively mild static fatigue effect, while the curve representing $A = 0.880$ and $B = -0.120$ is a relatively severe static fatigue effect. Also shown is the Wood curve (1951), which is the basis of the Madison curve (Figure 6.1), and which has often been used in design codes to represent static fatigue effects in joints (and all other

---

[10] Results presented and discussed here are brief extracts from comprehensive reliability studies varying parameters, such as type of joint (influences variability in strength), slope of rafters and truss spacing (Smith, 1985).

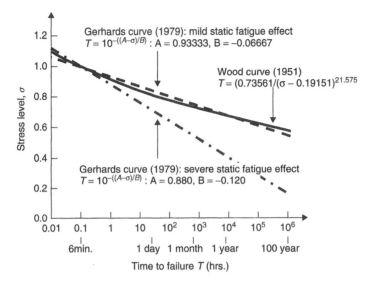

Figure 9.19 Problem 6: Gerhards curves for σ versus T adopted as basis of simulations, and comparison with Wood curve.

structural wood components). It can be seen that the Wood curve is quite mild, and approximately matches the Gerhards $A = 0.93333$ and $B = -0.06667$ curve. It is to be expected that in practice damage accumulation rates at a certain stress level σ will vary between components, either randomly or in some systematic way proportional to strength/quality. A key question that arises during reliability analysis, or indeed any other application of damage models, regards the influence inherent variability in rates at which components accumulate damage will have on residual strength and times to failure. Based on Gerhards model, Table 9.3 and Figure 9.20 show a range possible scenarios for variation in B (and indirectly A). These encompass either random variability unrelated to joint strength/material quality, and variation related to joint strength/material quality. Each of the listed possibilities has been suggested in the literature (see Chapters 6 and 8).

Following from the above, predictions have been generated to show the likely extent to which choice of damage model parameters (constants A and B) can influence allowable truss spans. Many such results exist, and Figure 9.21 gives an illustrative example only. It can be seen that there is very strong sensitivity to selection of parameters that

Table 9.3 Problem 6: Values of model constants used in simulations

Case	B — mean value	CoV (B)	Distribution
1	−.06667	0.0	*
2	−.06667	0.1	B normally distributed*
3	−.06667	0.2	B normally distributed*
4	−.06667	N/A	See Fig. 9.20
5	−.06667	N/A	See Fig. 9.20

*$A = 1.0 + B$.

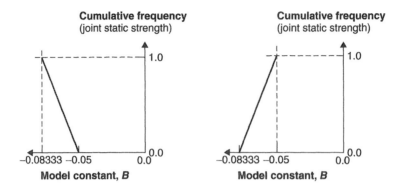

CASE 4:
Severity of static fatigue effect increases
as strength (quality) of joint increases

CASE 5:
Severity of static fatigue effect decrease
as strength (quality) of joint increases

Figure 9.20   Problem 6: Assignment of model constant B for Cases 4 and 5.

influence damage accumulation modelling. Taking Case 1 (no variability in B) as the
control, the only assumption that leads to liberal spans is Case 4 (increased rate of
damage accumulation if strength level/material quality increases). The most conservative prediction, i.e. lowest allowable spans, is associated with Case 3 (high level of
random variability in rate of damage accumulation). As there is no reason for suppos-
ing Case 1 is the most accurate, it should not be supposed that any other predictions
are necessarily unrealistic. What is shown by this analysis is that choices for parameters such as the target probability of failure (which relates to the level of structural
reliability), and the target design life are decisions of secondary importance. Although
perhaps counter intuitive, this is clearly so.

As discussed in Section 2.2, every wood component contains pre-existing (so-
called inherent) damage that initiated when the wood was part of a standing tree.
In Section 8.4 it was discussed that fracture mechanics principles imply that damage rates are strongly and positively correlated with strength level (wood quality),
implying that the most liberal situation, Case 4, is not inappropriate. If Case 4 is
realistic, it is clear that most modelling strategies will tend to be excessively conservative. In general, the situation that seems most likely is that both the strength
level (= quality $\propto$ amount of wood substance = porosity) and semi-randomised inherent damage strongly influence the rate of damage accumulation/time to failure.[11] Thus,
none of the above mentioned strategies for introducing variability in damage accumulation rates in models may be the optimal approach, i.e. Case 4 (B that increases
with any increase in strength level) in conjunction with randomised inherent damage.
This approach needs to be investigated. In the short term, as seen from Figure 9.21,
Case 1 solutions tend to be slightly but not excessively conservative relative to Case 4
solutions, likely making them a good choice.

---

[11] Strength level and inherent damage both also strongly influence initial strength (as it might be assessed
in a static load test).

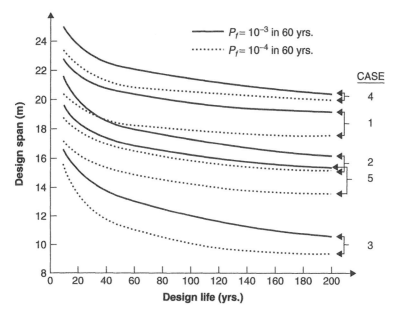

Figure 9.21   Problem 6: Effect of design life, damage model parameters and probability of failure P_f on design span of trusses.

This problem demonstrates that developers and users of damage models need to be fully cognisant of the underlying physical processes and how models are calibrated. In the absence of other guidance, an appropriate course of action is likely to be assumption of zero variability in damage parameters within reliability simulations (Case 1).

9.8   Problem 7: Fatigue in Bridge Stringers

This problem illustrates prediction of the number of load pulses to cause failure under High Cycle Fatigue (HCF) conditions. The system considered is a single-track short span railroad bridge having five parallel stringers (Figure 9.22). Modulus of elasticity E_i and moment capacity $M_{R,i}$ of each stringer is given in Table 9.4. The bridge is stiffened transversely by the rail ties, and this stiffening is sufficient to enforce equal stringer deflections under load.[12] Live load is from wheels of type EDM GP-38 locomotives and 263 k freight cars, each of which exerts maximum axle loads $F_{axle} = 292$ kN. Dead load is relatively small, and its effect is neglected here. For analysis stringers are presumed simply supported at each end, with strength controlled by the bending moment at mid-span. The force pulses produced by axles crossing the bridge are as shown in Figure 9.23. Peak stress level (moment at mid-span) for the ith

---

[12] Assuming flexural deformation dominates, the load carried by the ith stringer is proportional to $\dfrac{E I_i}{\sum_{i=1}^{5} E I_i}$, where $E I_i$ is the flexural rigidity of the ith stringer. As I is constant for all stringers, this reduces to $\dfrac{E_i}{\sum_{i=1}^{5} E_i}$.

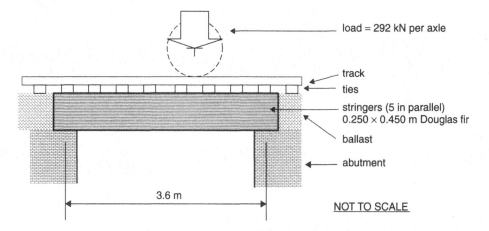

Figure 9.22   Problem 7: Arrangement for short span train bridge.

Table 9.4   Problem 7: Stringer data

Stringer	E_i (GPa)	$M_{R,i}$ (kNm)	$\dfrac{E_i}{\sum_{i=1}^{5} E_i}$	σ_i^*
1	5.8	180	0.153	0.224
2	7.6	145	0.201	0.364
3	11.0	210	0.291	0.364
4	7.9	140	0.209	0.392
5	5.5	140	0.146	0.273

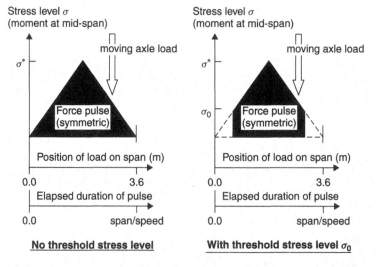

Figure 9.23   Problem 7: Definition of force pulses for calculation of fatigue damage.

stringer is:

$$\sigma_i^* = \frac{F_{axle}\, LE_i}{4M_{R,i}\displaystyle\sum_{i=1}^{5} E_i} = \frac{262.8E_i}{M_{R,i}\displaystyle\sum_{i=1}^{5} E_i} \qquad (9.12)$$

where $L = 3.6$ m is the span. Values of σ_i^* are given in Table 9.4. As shown in Figure 9.23, the duration of the force pulse is linearly proportional to the speed of the train.

Based on the above problem definition, the number of axle passes to failure is predicted for each stringer using the general form of the empirical damage accumulation model in Equation (8.33):

$$\mathrm{d}D = \sum_{i=1}^{I}\left(\int_0^{t_i} 10^{(A-\sigma_i)/B}\, \mathrm{d}t_i + \frac{1}{N_i}\right) \quad \text{if } \sigma_i > \sigma_{o,i} = \sigma_{o,0}(1 - D_{i-1})^C$$

$$\mathrm{d}D = 0 \qquad\qquad\qquad \text{if } \sigma_i \le \sigma_{o,i} = \sigma_{o,0}(1 - D_{i-1})^C \quad (9.13)$$

where D is the damage index ($0 \le D \le 1 =$ failure), t_i is the period for the ith load pulse, N_i is the pure fatigue life for the peak stress level associated with the ith load pulse, σ_i is the stress ratio at time t ($0 \le t \le t_i$), $\sigma_{o,i}$ is the threshold stress level for the ith load pulse, $\sigma_{o,0}$ is the threshold stress level when $D = 0$, D_{i-1} is the damage level at the end of the $(i\text{-}1)$th load pulse, and A, B and C are calibration constants. As discussed in Chapter 8, the first term in the expression for $\mathrm{d}D$ accounts for static fatigue effects, and is proportional to the duration of a force pulse. Obviously, its damage effect accumulates more rapidly the slower the speed of the trains. It follows from Equation (9.13) that the increment in $\mathrm{d}D$ per axle pass due to static fatigue is:

$$\mathrm{d}D_{static} = \frac{-BL(10^{(A-\sigma^*)/B} - 10^{(A-\sigma_0)/B})}{2.3026s\sigma^*} \qquad (9.14)$$

where s is the speed of the train, and other parameters are as previously defined. For the purposes of this problem is presumed that $A = 0.880$ and $B = -0.120$, based on the literature (Chapter 6).

The second term in Equation (9.13) accounts for effects of the rapid loading and unloading rates associated with very short duration load pulses, i.e. cyclic fatigue effect. The increment in $\mathrm{d}D$ per axle pass due to cyclic fatigue is taken to be:

$$\mathrm{d}D_{cyclic} = 10^{-12.5(1-\sigma^*)} \qquad (9.15)$$

This corresponds to the stress level versus fatigue life relationship in Figure 9.24, based on data for structural material in wet condition (Tsai and Ansell, 1990). Strictly, any expression for $\mathrm{d}D_{\text{cyclic}}$ should be a function of loading frequency (train speed) and the waveform (see Section 6.2.2), but such refinement is excluded here as that would not influence predictions, as shown below.

Table 9.5 gives predicted numbers of axle passes to fail each stringer if only the cyclic fatigue effect is considered. Results represent the extreme case of: a zero threshold stress level throughout loading ($\sigma_{o,0} = 0.0$), and a non-zero and constant threshold

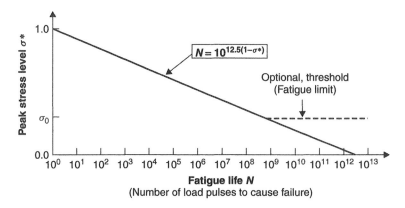

Figure 9.24   Problem 7: Assumed peak stress level vs. pure fatigue life ($\sigma^* - N$) relationship.

Table 9.5   Problem 7: Results — cyclic fatigue effect only

Stringer	Number of axle passes to failure	
	$\sigma_{o,0} = 0.0$	$\sigma_{o,0} = 0.3; C = 0.0$
1	$10^{9.7}$	∞
2	$10^{8.0}$	$10^{8.0}$
3	$10^{8.0}$	$10^{8.0}$
4	$10^{7.6}$	$10^{7.6}$
5	$10^{9.1}$	∞

stress level throughout loading ($\sigma_{o,0} = 0.3$, $C = 0.0$).[13] As can be seen, the number of axle passes to cause failure differs quite substantially between stringers. The bridge as a whole is predicted to failure after $10^{7.6}$ axle passes (stringer 4 governs). Table 9.6 gives predicted numbers of axle passes to fail each stringer if only the static fatigue effect is considered. Results are for two train speeds, and the extreme cases of $\sigma_{o,0} = 0.0$; and $\sigma_{o,0} = 0.3$, $C = 0.0$. For the various stringers, the predicted number of axle passes to cause failure is about two to three orders of magnitude less than for cyclic fatigue alone. Clearly, the static fatigue effect dominates the outcome of predictions. As shown, train speed and existence of a threshold stress level have secondary influences on results. The bridge as a whole is predicted to fail after about 10^5 axle passes (again stringer 4 governs). To a good approximation, predictions for full implementation of Equation (9.13), i.e. $dD = dD_{\text{static}} + dD_{\text{cyclic}}$, are the same as those in Table 9.6, because $dD_{\text{static}} \gg dD_{\text{cyclic}}$.

As mentioned in a footnote to the above, results for models with a threshold stress level that degrades as damage accumulates ($C > 0.0$) are intermediate between those obtained assuming $\sigma_{o,0} = 0.0$; and $\sigma_{o,0} = 0.3$, $C = 0.0$. To illustrate, if $\sigma_{o,0} = 0.3$ and $C = 1.0$ the number of axle passes to fail the bridge would be $10^{5.34}$, instead of $10^{5.31}$ when $\sigma_{o,0} = 0.0$.

---

[13] Results for models with a threshold stress level that degrades as damage increases, $C > 0.0$, are intermediate between the two extremes.

Table 9.6 Problem 7: Results—static fatigue effect only[a]

Stringer	Number of axle passes to failure			
	Speed = 30 km/hr		Speed = 60 km/hr	
	$\sigma_{o,0} = 0.0$	$\sigma_{o,0} = 0.3$; $C = 0.0$	$\sigma_{o,0} = 0.0$	$\sigma_{o,0} = 0.3$; $C = 0.0$
1	$10^{6.5}$	∞	$10^{6.8}$	∞
2	$10^{5.5}$	$10^{5.7}$	$10^{5.8}$	$10^{6.0}$
3	$10^{5.5}$	$10^{5.7}$	$10^{5.8}$	$10^{6.0}$
4	$10^{5.3}$	$10^{5.4}$	$10^{5.6}$	$10^{5.7}$
5	$10^{6.1}$	∞	$10^{6.4}$	∞

[a]To an engineering level approximation, results are the same as the above if cyclic and static fatigue effects are combined.

Readers should not generalise the outcome of this particular problem regarding relative importance of dD_{static} and dD_{cyclic} contributions to dD, and therefore about the life expectancy of structural systems. Had the static fatigue effect been taken to be much milder, or the cyclic fatigue effect more acute, both terms would have been significant. Relative importance of cyclic fatigue or static fatigue effects on the damage accumulation processes is a situation specific outcome.

This problem illustrates that dD_{static} and dD_{cyclic} type terms in damage index models have distinct purposes and mimic different components of the physical process. Recognition of this needs to carry through into development of models, interpretation of data, calibration of models and implementation of models.

9.9 Problem 8: Cyclic Fatigue in Nailed Joints

Most mechanical joints in structural wood systems employ metal fasteners, and in many cases, it is the strength characteristics of wood in the vicinity of fasteners that determines the failure mechanism irrespective of the loading regime. This is the situation for joints with laterally loaded dowel type fasteners if the fasteners are stocky and/or the members relatively thin (see Problem 1, Section 9.2). However, this is not the case if fasteners are slender. This problem considers Low Cycle Fatigue (LCF) behaviour of laterally loaded wood-to-wood joints made with slender nails.

LCF tests have been performed at the University of New Brunswick on single-shear one-nail joints in air-dried white ash (density 600 kg/m^3). Joints were made with common (steel wire) nails 3.67 mm in diameter and 76 mm in length. The grain of the wood was oriented so nails loaded both members parallel to grain. Specimens had the dimensions shown in Figure 9.25. Matched tests measured static strength of the joints, cyclic load-embedment characteristics for nails bearing on members, and fatigue behaviour of isolated nails. All cyclic tests employed fully reversed load applied under displacement control, with sinusoidal waveform and loading frequency of 0.5 Hz. Repetitive displacements were applied until failure of joints or isolated nails, with amplitudes selected to produce smooth displacement versus fatigue life ($\delta - N$) curves. In the nonlinear regime, but prior to ultimate failure, the nail in a joint took on the distinctive deformed shape shown in Figure 9.26(a). This shape is associated with the

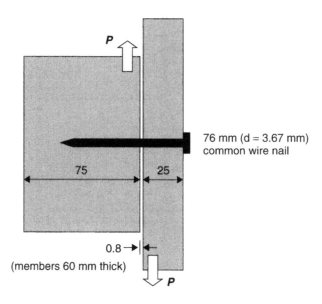

Figure 9.25   Problem 8: Single-shear one-nail joint specimen.

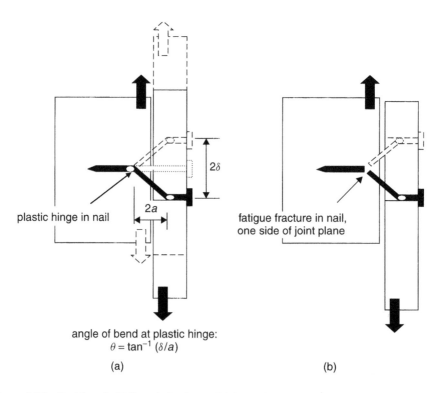

Figure 9.26   Problem 8: Failure behaviour of joints with cyclically laterally loaded slender nails. (a) Mechanism in the nonlinear regime, but prior to ultimate failure; slip amplitude 2δ (movements $\pm\delta$); (b) mechanism at ultimate failure.

formation of a hinge either side of the joint plane, and pseudo-plastic crushing of wood immediately beneath the nail at locations between the hinges. Although nail movement in the wood can induce cracking parallel to grain adjacent to the joint plane, crack development is stable up to and including ultimate failure. Hinges were measured to lie 7.13 mm (CoV 11%) below the surface of members (accounting for the inter-member gap of 0.8 mm, $2a = 15.06$ mm in Figure 9.26(a)). Thus, neglecting elastic deformation, the angle of bend in hinges $\theta = \tan^{-1} (2\delta/15.06)$, where 2δ is the slip amplitude within repetitive load cycles. Ultimate failure was due to nail fracture, as illustrated in Figure 9.26(b). Types of nail fracture observed were ductile necking and brittle planar failures. A typical load versus time response for a joint that exhibited ductile necking failure in the nail is shown in Figure 9.27. This type of failure occurred under high amplitude cyclic slip. A typical load versus time response for a joint that exhibited brittle planar nail failure is shown in Figure 9.28. Such failures occurred under relatively low amplitude cyclic slip. As seen in both Figures 9.27 and 9.28, even initial deformation cycles can degrade load-carrying capability of a joint. In cases where the nail failure is ductile, this reflects accumulation of residual strain under displacement control, rather than a significant compromise of residual strength. There is always a marked drop in residual capacity as the number of load cycles approaches the fatigue life at any slip amplitude.

Based on test observations, it is clear that LCF in joints with slender nails mainly results from strain in nails, and failure is related to the extent of rotation of hinges in nails. It follows that if the rotation versus fatigue life relationship for hinges in nails is known, then the fatigue life of joints can be predicted. Figure 9.29 compares θ versus fatigue life N of isolated nails with θ for nails in joints. (θ in joints is estimated by the approach indicated in Figure 9.26.) In general agreement is quite good, despite a tendency for behaviour of isolated nails to underestimate the hinge

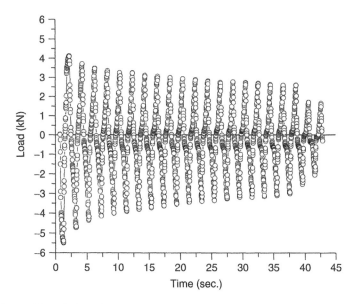

Figure 9.27 Problem 8: Load versus time response of a fatigued joint exhibiting ductile nail failure.

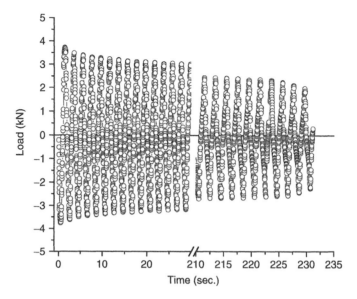

Figure 9.28   Problem 8: Load versus time response of a fatigued joint exhibiting brittle planar nail failure.

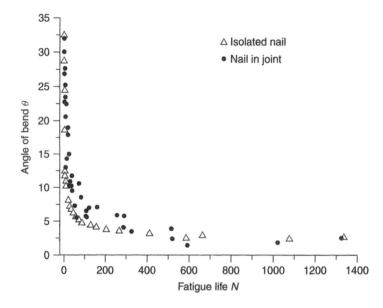

Figure 9.29   Problem 8: Plastic hinge angles in isolated nail tests and joint tests.

angle for nails in joints. In absolute terms the discrepancy is quite small (2° to 3°) and within the bounds of experimental error. The error is attributable to difficulty in separating inelastic/plastic and elastic components of displacement. Although not discussed in detail here, the locations of hinges in nails can be calculated accurately based on plastic theory of joints (Smith *et al.*, 2002). Such analysis assumes both nails and wood are ideal rigid-plastic materials (Johansen, 1949). Necessary information for model predictions of joint fatigue life is load-embedment response of wood members, fatigue data for isolated nails and joint geometry. Such limited information enables predictions of fatigue life of myriad joints without resort to testing. It is, of course, wise to carry out some confirmatory testing on whole joints.

This problem illustrates that joints, and therefore structural wood systems, are not always critically influenced by brittle processes associated with fracture or fatigue of the wood components. It is necessary, therefore, to be more discriminating than to simply extrapolate from behaviour of wood to imply behaviour of systems. Unfortunately, such an extrapolation is embodied in the current generation of engineering design codes!

9.10   References

Aicher, S., Schmidt, S. and Brunold S. (1995) 'Design of timber beams with holes by means of fracture mechanics', Paper 28-19-4, *Proceedings of CIB-Working Commission 18: Timber Structures, Meeting 28*, International Council for Research and Innovation in Building and Construction, Rotterdam, The Netherlands.

Bray, D.E. and Stanley, R.K. (1989) Nondestructive Evaluation, McGraw-Hill, New York, NY, USA.

Foschi, R.O., Folz, B.R. and Yao, F.Z. (1989) 'Reliability-based design of wood structures', Structural Research Series Report No. 34, University of British Columbia, Vancouver, BC, Canada.

Gerhards, C.C. (1979) 'Time-related effects of loading on wood strength: a linear cumulative damage theory', *Wood Science*, **11**(3): 139–144.

Hanh, G.J. and Shapiro, S.S. (1967) Statistical Models in Engineering, John Wiley and Sons, New York, NY, USA.

HMSO (1955) 'The machine boring of wood', Forest Products Research Bulletin No. 35, Her Majesty's Stationary Office, London, UK.

Johansen, K.W. (1949) 'Theory of timber connectors', Publication No. 9. International Association for Bridge and Structural Engineering, Bern, Switzerland: 249–262.

Kharouf, N. (2001) 'Post-elastic behavior of bolted connections in wood', PhD thesis, McGill University, Montreal, QC, Canada.

Landelius, J. (1989) 'Finite area method', Report TVSM-5043, Lund Institute of Technology, Division of Structural Mechanics, University of Lund, Lund, Sweden.

Sanchez, O. (2002) 'Performance study of frp-glulam bridge girders', MS Thesis, University of Maine, Orono, ME, USA.

Smith, I. (1985) 'Methods for calibrating design factors in partial coefficients limit states design codes for structural timberwork: with special reference to mechanical timber joints', Research Report 1/85, Timber Research and Development Association, High Wycombe, UK.

Smith, I. and Springer, G. (1993) 'Consideration of Gustafsson's proposed Eurocode 5 failure criterion for notched timber beams', *Canadian Journal of Civil Engineering*, **20**(6): 1030–1036.

Smith, I., Chui, Y.H. and Hu, L-J. (1996a) 'Reliability analysis for critical reaction forces for lumber members with an end notch', *Canadian Journal of Civil Engineering*, **23**(1): 202–210.

Smith, I., Gong, M. and Foliente, G. (2002) 'Predicting cyclic fatigue behaviour of laterally loaded nailed timber joints', *Proceedings of World Conference on Timber Engineering*, MARA University of Technology, Shah Alam, Malaysia: 2.472–478.

Smith, I., Tan, D. and Chui, Y.H. (1996b) 'Critical reaction forces for notched timber members', *Proceedings International Wood Engineering Conference*, Louisiana State University, Baton Rouge, USA, 3: 460–466.

Tan, D. and Smith, I. (1998) 'Effect of notch and radius on the strength perpendicular to grain of spruce,' *Proceedings 5th World Conference on Timber Engineering*, Lausanne, Switzerland, 2: 748–749.

Tan, D. and Smith, I. (1999) 'Failure in-the-row model for bolted timber connections', *ASCE Journal of Structural Engineering*, **125**(7): 713–718.

Tsai, K.T. and Ansell, M.P. (1990) 'The fatigue properties of wood in flexure', *Journal of Material Science*, **25**: 865–878.

Valentin, G.H., Bostrom, L., Gustafsson, P.J., Ranta-Maunus, A. and Gowda, S. (1991) 'Application of fracture mechanics to timber structures', RILEM State-of-the-art Report, Technical Research Centre of Finland, Research Notes 1262, Espoo, Finland.

Wood, L. (1951) 'Relation of strength of wood to duration of load', Report No. 1916, US Forest Products Laboratory, Madison, WI, USA.

Index

accelerated test methods   124
accumulated deformation   129–30, 143
average stress criterion   205
Axiom of Uniform Stress   19–20
axes of symmetry   37–8, 45, 47, 52–3, 59
axis of growth   100

bacterial damage   21
bark   9, 16
Barrett and Foschi type   139, 179–81, 183,
    192–3
basic density   22
bound water   23
bridging model   164–6
Burger Body 4-element dash-pot and spring
    model   55–7

cambium   9, 16–17
cell functions   15
cell walls   13–14, 39–41
    middle lamella   13
    middle layer   13
    outer layer   13
    primary wall   13
    secondary wall   13
cellulose   12
cellulose tension hypothesis   10
chemical components   12–13
chemical treatment, effect on fatigue
    properties   137
clear wood
    constitutive equations   43–7
    in-elastic response   50–2
    mechanical properties   27
    modelling   43–58
    stiffness   49
    strength criteria   52–3

cohesive zone   89
composites (wood with glue type)   124, 134,
    177, 210–13
compression failure   19
compressive stresses   10
constitutive equations
    of clear wood   43–7
    of massive wood   59
continuum damage mechanics   94–5
Cook–Gordon crack stopping   93, 104,
    105
Cowin strength criteria   53
crack bridging   104
crack growth   68–71
Crack Mouth Opening Displacement (CMOD)
    114–15
crack opening displacement   90–1, 191
crack resistance   71–6
Crack Tip Opening Displacement (CTOD)
    90–1
creep *see* rheological behaviour
creep rupture   54
    see also static fatigue
critical crack analysis   210–13
critical crack length   80–1
critical strain energy release rate   71–6,
    81–2, 82–3
critical stress intensity factor   101
cross grain   21
crystalline structure   12
cyclic fatigue
    experiment studies   123–4
    in massive wood   135–8, 223–7
    models   177–8, 189, 191, 194, 219–23,
        223–7

Fracture and Fatigue in Wood   I. Smith, E. Landis and M. Gong
© 2003 John Wiley & Sons, Ltd   ISBN: 0-471-48708-2 (HB)

Lightning Source UK Ltd.
Milton Keynes UK
UKHW051723311019
352550UK00002BB/46/P